CONVECTION

RADIATION

CONDUCTION

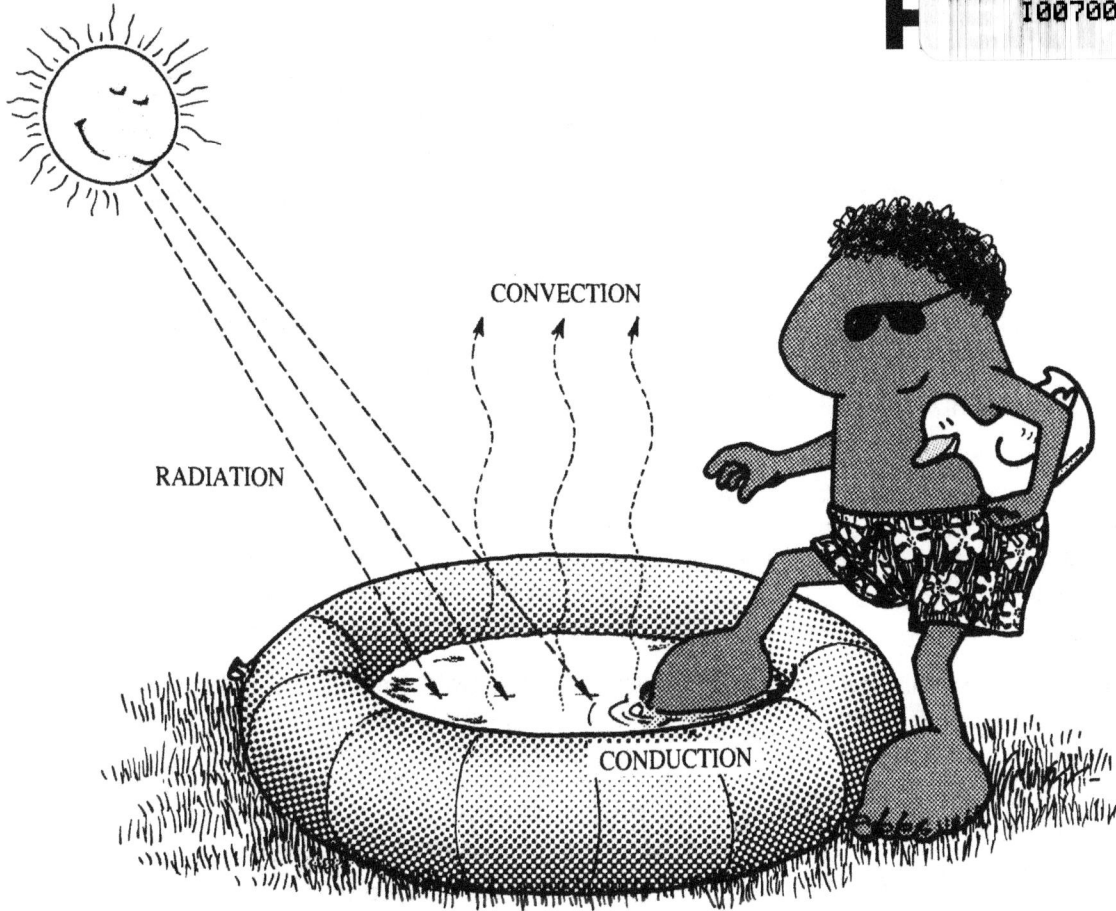

TASK CARD SERIES

Conceived and
written by
Ron Marson
Illustrated by
Peg Marson

TOPS LEARNING SYSTEMS

342 S Plumas Street
Willows CA 95988

www.topsscience.org

WHAT CAN YOU COPY?

Dear Educator,

Please honor our copyright restrictions. We offer liberal options and guidelines below with the intention of balancing your needs with ours. When you buy these labs and use them for your own teaching, you sustain our work. If you "loan" or circulate copies to others without compensating TOPS, you squeeze us financially, and make it harder for our small non-profit to survive. Our well-being rests in your hands. Please help us keep our low-cost, creative lessons available to students everywhere. Thank you!

PURCHASE, ROYALTY and LICENSE OPTIONS

TEACHERS, HOMESCHOOLERS, LIBRARIES:

We do all we can to keep our prices low. Like any business, we have ongoing expenses to meet. We trust our users to observe the terms of our copyright restrictions. While we prefer that all users purchase their own TOPS labs, we accept that real-life situations sometimes call for flexibility.

Reselling, trading, or loaning our materials is prohibited unless one or both parties contribute an Honor System Royalty as fair compensation for value received. We suggest the following amounts – let your conscience be your guide.

HONOR SYSTEM ROYALTIES: If making copies from a library, or sharing copies with colleagues, please calculate their value at 50 cents per lesson, or 25 cents for homeschoolers. This contribution may be made at our website or by mail (addresses at the bottom of this page). Any additional tax-deductible contributions to make our ongoing work possible will be accepted gratefully and used well.

Please follow through promptly on your good intentions. Stay legal, and do the right thing.

SCHOOLS, DISTRICTS, and HOMESCHOOL CO-OPS:

PURCHASE Option: Order a book in quantities equal to the number of target classrooms or homes, and receive quantity discounts. If you order 5 books or downloads, for example, then you have unrestricted use of this curriculum for any 5 classrooms or families per year for the life of your institution or co-op.

2-9 copies of any title: 90% of current catalog price + shipping.

10+ copies of any title: 80% of current catalog price + shipping.

ROYALTY/LICENSE Option: Purchase just one book or download *plus* photocopy or printing rights for a designated number of classrooms or families. If you pay for 5 additional Licenses, for example, then you have purchased reproduction rights for an entire book or download edition for any **6** classrooms or families per year for the life of your institution or co-op.

1-9 Licenses: 70% of current catalog price per designated classroom or home.

10+ Licenses: 60% of current catalog price per designated classroom or home.

WORKSHOPS and TEACHER TRAINING PROGRAMS:

We are grateful to all of you who spread the word about TOPS. Please limit copies to only those lessons you will be using, and collect all copyrighted materials afterward. No take-home copies, please. Copies of copies are strictly prohibited.

For licensing, honor system royalty payments, contact: **www.TOPScience.org**; or **TOPS Learning Systems 342 S Plumas St, Willows CA 95988**; or inquire at **customerservice@topscience.org**

ISBN 978 - 0 - 941008 - 85 - 3

CONTENTS

A TOPS Model for Effective Science Teaching...

If science were only a set of explanations and a collection of facts, you could teach it with blackboard and chalk. You could assign students to read chapters and answer the questions that followed. Good students would take notes, read the text, turn in assignments, then give you all this information back again on a final exam. Science is traditionally taught in this manner. Everybody learns the same body of information at the same time. Class togetherness is preserved.

But science is more than this.

Science is also process — a dynamic interaction of rational inquiry and creative play. Scientists probe, poke, handle, observe, question, think up theories, test ideas, jump to conclusions, make mistakes, revise, synthesize, communicate, disagree and discover. Students can understand science as process only if they are free to think and act like scientists, in a classroom that recognizes and honors individual differences.

Science is *both* a traditional body of knowledge *and* an individualized process of creative inquiry. Science as process cannot ignore tradition. We stand on the shoulders of those who have gone before. If each generation reinvents the wheel, there is no time to discover the stars. Nor can traditional science continue to evolve and redefine itself without process. Science without this cutting edge of discovery is a static, dead thing.

Here is a teaching model that combines the best of both elements into one integrated whole. It is only a model. Like any scientific theory, it must give way over time to new and better ideas. We challenge you to incorporate this TOPS model into your own teaching practice. Change it and make it better so it works for you.

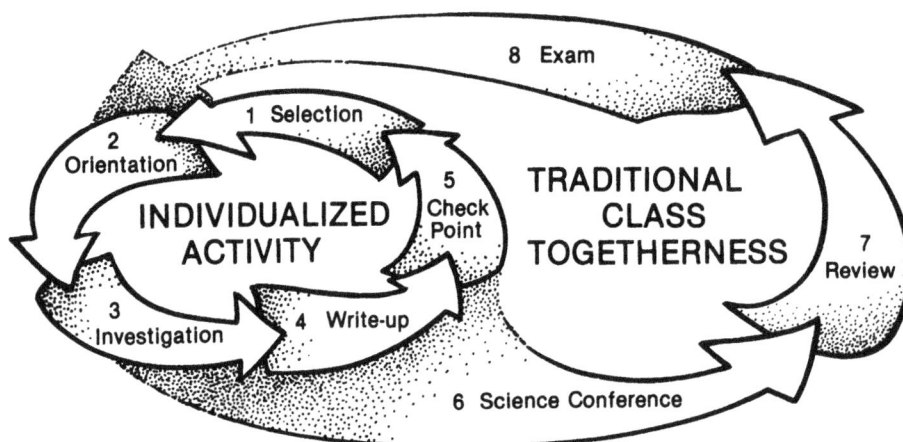

1. SELECTION

Doing TOPS is as easy as selecting the first task card and doing what it says, then the second, then the third, and so on. Working at their own pace, students fall into a natural routine that creates stability and order. They still have questions and problems, to be sure, but students know where they are and where they need to go.

Students generally select task cards in sequence because new concepts build on old ones in a specific order. There are, however, exceptions to this rule: students might *skip* a task that is not challenging; *repeat* a task with doubtful results; *add* a task of their own design to answer original "what would happen if" questions.

2. ORIENTATION

Many students will simply read a task card and immediately understand what to do. Others will require further verbal interpretation. Identify poor readers in your class. When they ask, "What does this mean?" they may be asking in reality, "Will you please read this card aloud?"

With such a diverse range of talent among students, how can you individualize activity and still hope to finish this module as a cohesive group? It's easy. By the time your most advanced students have completed all the task cards, including the enrichment series at the end, your slower students have at least completed the basic core curriculum. This core provides the common background so necessary for meaningful discussion, review and testing on a class basis.

3. INVESTIGATION

Students work through the task cards independently and cooperatively. They follow their own experimental strategies and help each other. You encourage this behavior by helping students only *after* they have tried to help themselves. As a resource person, you work to stay *out* of the center of attention, answering student questions rather than posing teacher questions.

When you need to speak to everyone at once, it is appropriate to interrupt individual task card activity and address the whole class, rather than repeat yourself over and over again. If you plan ahead, you'll find that most interruptions can fit into brief introductory remarks at the beginning of each new period.

4. WRITE-UP

Task cards ask students to explain the "how and why" of things. Write-ups are brief and to the point. Students may accelerate their pace through the task cards by writing these reports out of class.

Students may work alone or in cooperative lab groups. But each one must prepare an original write-up. These must be brought to the teacher for approval as soon as they are completed. Avoid dealing with too many write-ups near the end of the module, by enforcing this simple rule: each write-up must be approved *before* continuing on to the next task card.

5. CHECK POINT

The student and teacher evaluate each write-up together on a pass/no-pass basis. (Thus no time is wasted haggling over grades.) If the student has made reasonable effort consistent with individual ability, the write-up is checked off on a progress chart and included in the student's personal assignment folder or notebook kept on file in class.

Because the student is present when you evaluate, feedback is immediate and effective. A few seconds of this direct student-teacher interaction is surely more effective than 5 minutes worth of margin notes that students may or may not heed. Remember, you don't have to point out every error. Zero in on particulars. If reasonable effort has not been made, direct students to make specific improvements, and see you again for a follow-up check point.

A responsible lab assistant can double the amount of individual attention each student receives. If he or she is mature and respected by your students, have the assistant check the even-numbered write-ups while you check the odd ones. This will balance the work load and insure that all students receive equal treatment.

6. SCIENCE CONFERENCE

After individualized task card activity has ended, this is a time for students to come together, to discuss experimental results, to debate and draw conclusions. Slower students learn about the enrichment activities of faster students. Those who did original investigations, or made unusual discoveries, share this information with their peers, just like scientists at a real conference. This conference is open to films, newspaper articles and community speakers. It is a perfect time to consider the technological and social implications of the topic you are studying.

7. REVIEW

Does your school have an adopted science textbook? Do parts of your science syllabus still need to be covered? Now is the time to integrate other traditional science resources into your overall program. Your students already share a common background of hands-on lab work. With this shared base of experience, they can now read the text with greater understanding, think and problem-solve more successfully, communicate more effectively.

You might spend just a day on this step or an entire week. Finish with a review of key concepts in preparation for the final exam. Test questions in this module provide an excellent basis for discussion and study.

8. EXAM

Use any combination of the review/test questions, plus questions of your own, to determine how well students have mastered the concepts they've been learning. Those who finish your exam early might begin work on the first activity in the next new TOPS module.

Now that your class has completed a major TOPS learning cycle, it's time to start fresh with a brand new topic. Those who messed up and got behind don't need to stay there. Everyone begins the new topic on an equal footing. This frequent change of pace encourages your students to work hard, to enjoy what they learn, and thereby grow in scientific literacy.

Getting Ready

Here is a checklist of things to think about and preparations to make before your first lesson.

✔ Review the scope and sequence.

Take just a few minutes, right now, to page through each activity. Read each *Task Objective* at the top of the page and scan each task card underneath.

✔ Set aside appropriate class time.

Allow an average of perhaps 1 class period per lesson (more for younger students), plus time at the end of this module for discussion, review and testing. If your schedule doesn't allow this much science, consult the logic tree on page E to see which activities you can safely omit without breaking conceptual links between lessons.

✔ Number your task card masters.

The small number printed in the lower right corner of each task card shows its position within the series. If this ordering fits your schedule, copy each number into the blank parentheses directly above it at the top of the card. Be sure to use pencil; you may decide to revise, rearrange, add or omit task cards the next time you teach this module. Simply insert your own better ideas on 4 x 6 index cards wherever they fit best, and renumber your sequence. This allows your curriculum to adapt and grow as you do.

✔ Photocopy task cards for student use.

Decide how you want to distribute task cards among your students. Find a list of management options opposite the task card master numbered "cards 1-2." Then photocopy and collate classroom sets of task cards in the numbers you need, plus supplementary pages (if any) at the back of this module. The *Materials* list that accompanies each activity specifies when these supplementary pages are required, and how many to photocopy.

We allow you to photocopy all permissible materials, as long as you limit the distribution of copies you make to the students you personally teach. Encourage other teachers who want to use this module to purchase their own TOPS module. This supports TOPS financially, enabling us to continue publishing new modules for you.

✔ Collect needed materials.

Please see page D, opposite, for details.

✔ Organize a way to track assignments.

Keep student work on file in class. If you lack a file cabinet, a box with a brick will serve. File folders or notebooks both make suitable assignment organizers. Students will feel a sense of accomplishment as they see their folders grow heavy, or their notebooks fill, with completed assignments. Since all papers stay together, reference and review are facilitated.

Ask students to number a sheet of paper 1, 2, 3..., to the total number of task cards in this module, then tape it inside the front cover of their folders or notebooks. Track individual progress through this module (and future modules) by initialing lesson numbers as students complete each assignment.

✔ Review safety procedures.

In our litigation-conscious society, we find that publishers are often more committed to protecting themselves from liability suits than protecting students from physical hazards. Lab instructions are too often filled with spurious advisories, cautions and warnings that desensitize students to safety in general. If we cry "Wolf!" too often, real warnings of present danger may go unheeded.

At TOPS we endeavor to use good sense in deciding what students already know (don't stab yourself in the eye) and what they should be told (don't look directly at the sun.) Scissors and pins could be dangerous in the hands of unsupervised children. Nor can this curriculum anticipate irresponsible behavior or negligence. As the teacher, it is ultimately your responsibility to see that common-sense safety rules are followed; it is your students' responsibility to respect and protect themselves and each other.

Begin with these **BASIC SAFETY RULES**

ALLERGIES: Some people are dangerously allergic to peanuts, used in activity 20, so send sensitive kids to an alternate location. Traces of oil on fingers or in smoke may pose a risk. Wipe desks with soapy water when finished.

EYE PROTECTION: Wear safety goggles when heating liquids or solids to high temperatures.

POISONS: Never taste anything unless told to do so.

FIRE: Keep hair and clothing away from flames. Aim test tubes away from your face, and your neighbor's, when heating.

GAS: Light the match first, before turning on the gas.

✔ Communicate your grading expectations.

Whatever your grading philosophy, your students need to understand how they will be assessed. Here is a scheme that counts individual effort, attitude and overall achievement. We think these 3 components deserve equal weight:

• <u>Pace</u> (effort): Tally the number of check points and extra credit experiments you have initialed for each student. Low-ability students should be able to keep pace with gifted students, since write-ups are evaluated relative to individual performance standards on a pass/no-pass basis. Students with absences or those who tend to work slowly might assign themselves more homework out of class.

• <u>Participation</u> (attitude): This is a subjective grade, assigned to measure personal initiative and responsibility. Active participators who work to capacity receive high marks. Inactive onlookers who waste time in class and copy the results of others receive low marks.

• <u>Exam</u> (achievement): Activities point toward generalizations that provide a basis for hypothesizing and predicting. Review/ Test questions beginning on page G will help you assess whether students understand relevant theory and can apply it in a predictive way.

Gathering Materials

Listed below is everything you'll need to teach this module. Buy what you don't already have from your local supermarket, drugstore or hardware store. Ask students to bring recycled materials from home.

Keep this classification key in mind as you review what's needed.

general on-the-shelf materials:	special in-a-box materials:
Normal type suggests that these materials are common. Keep these basics on shelves or in drawers that are accessible to your students. The next TOPS module you teach will likely utilize many of these same materials.	*Italic type suggests that these materials are unusual. Keep these specialty items in a separate box. After you finish teaching this module, label the box for storage and put it away, ready to use again.*
(substituted materials):	*optional materials:
Parentheses enclosing any item suggests a ready substitute. These alternatives may work just as well as the original. Don't be afraid to improvise, to make do with what you have.	An asterisk sets these items apart. They are nice to have, but you can easily live without them. They are probably not worth an extra trip to the store, unless you are gathering other materials as well.

Everything is listed in order of first use. Start gathering at the top of this list and work down. (The teaching notes may occasionally suggest additional *Extensions*. Supplies for these optional experiments are listed neither here nor under *Materials*. Read the extension itself to determine what new items, if any, are required.)

Quantities depend on how many students you have, how you organize them into activity groups, and how you teach. Decide which of these 3 estimates best applies to you, then adjust quantities up or down as necessary:

$Q_1 / Q_2 / Q_3$

- **Single Student:** Enough for 1 student to do all the experiments.
- **Individualized Approach:** Enough for 30 students informally working in pairs, all self-paced.
- **Traditional Approach:** Enough for 30 students, organized into pairs, all doing the same lesson.

KEY:	general on-the-shelf materials (substituted materials)	special in-a-box materials *optional materials

	general on-the-shelf		special in-a-box
1/1/1	rolls of iron, copper and aluminum wire of roughly equal thickness, perhaps 20 gauge	1/10/10	toothpicks
1/5/10	*wire cutters	2/10/20	tin can tops, about 15 ounce size
1/10/10	candles with drip catchers	1/5/10	tin cans, about 15 ounce size
1/10/10	books of matches	2/10/20	pennies
1/10/10	glass microscope slides	1/1/1	roll aluminum foil
1/10/10	scissors	1/2/2	rolls clear tape
3/20/30	*thin, recyclable, aluminum pie tins*	1/10/10	laboratory thermometers
1/2/2	trays of ice cubes	1/3/3	sheets black paper (or color white paper with black crayon or marking pen)
1/10/10	plastic sandwich bags	1/1/1	hot plate and teapot to heat water (or use Bunsen burners, Pyrex beakers and ring stands)
1/2/2	rolls masking tape		
1/1/1	dropper bottle of blue food coloring		
1/1/1	source of hot and cold water	3/30/30	corks to fit test tubes (lumps of oil-based clay)
1/10/10	Bunsen burners or alcohol lamps — other heating sources may be substituted in all experiments except activity 9	1/1/1	jar of sand
		1/10/10	graduated cylinders, 100 mL capacity
		2/20/20	small styrofoam cups — 150 mL minimum capacity
2/10/20	pint jars	1/5/10	*large plastic milk jugs cut to half size, or equivalent*
3/20/30	baby food jars	1/5/10	*graduated cylinders, 1000 mL capacity (quart or liter jars)
2/20/20	index cards		
1/1/1	spool of thread	1/10/10	hand calculators
3/30/30	test tubes	.1/.3/.8	*kilograms washers, bolts or other small iron objects*
1/1/1	box steel wool	1/5/10	gram balances
1/1/1	box paper clips	1/1/1	small container flour
2/20/20	clothespins	1/1/1	*bag roasted peanuts*
		1/10/10	straight pins

D

Sequencing Task Cards

This logic tree shows how all the task cards in this module tie together. In general, students begin at the trunk of the tree and work up through the related branches. As the diagram suggests, the way to upper level activities leads up from lower level activities.

At the teacher's discretion, certain activities can be omitted or sequences changed to meet specific class needs. The only activities that must be completed in sequence are indicated by leaves that open *vertically* into the ones above them. In these cases the lower activity is a prerequisite to the upper.

When possible, students should complete the task cards in the same sequence as numbered. If time is short, however, or certain students need to catch up, you can use the logic tree to identify concept-related *horizontal* activities. Some of these might be omitted since they serve only to reinforce learned concepts rather than introduce new ones.

On the other hand, if students complete all the activities at a certain horizontal concept level, then experience difficulty at the next higher level, you might go back down the logic tree to have students repeat specific key activities for greater reinforcement.

For whatever reason, when you wish to make sequence changes, you'll find this logic tree a valuable reference. Parentheses in the upper right corner of each task card allow you total flexibility They are left blank so you can pencil in sequence numbers of your own choosing.

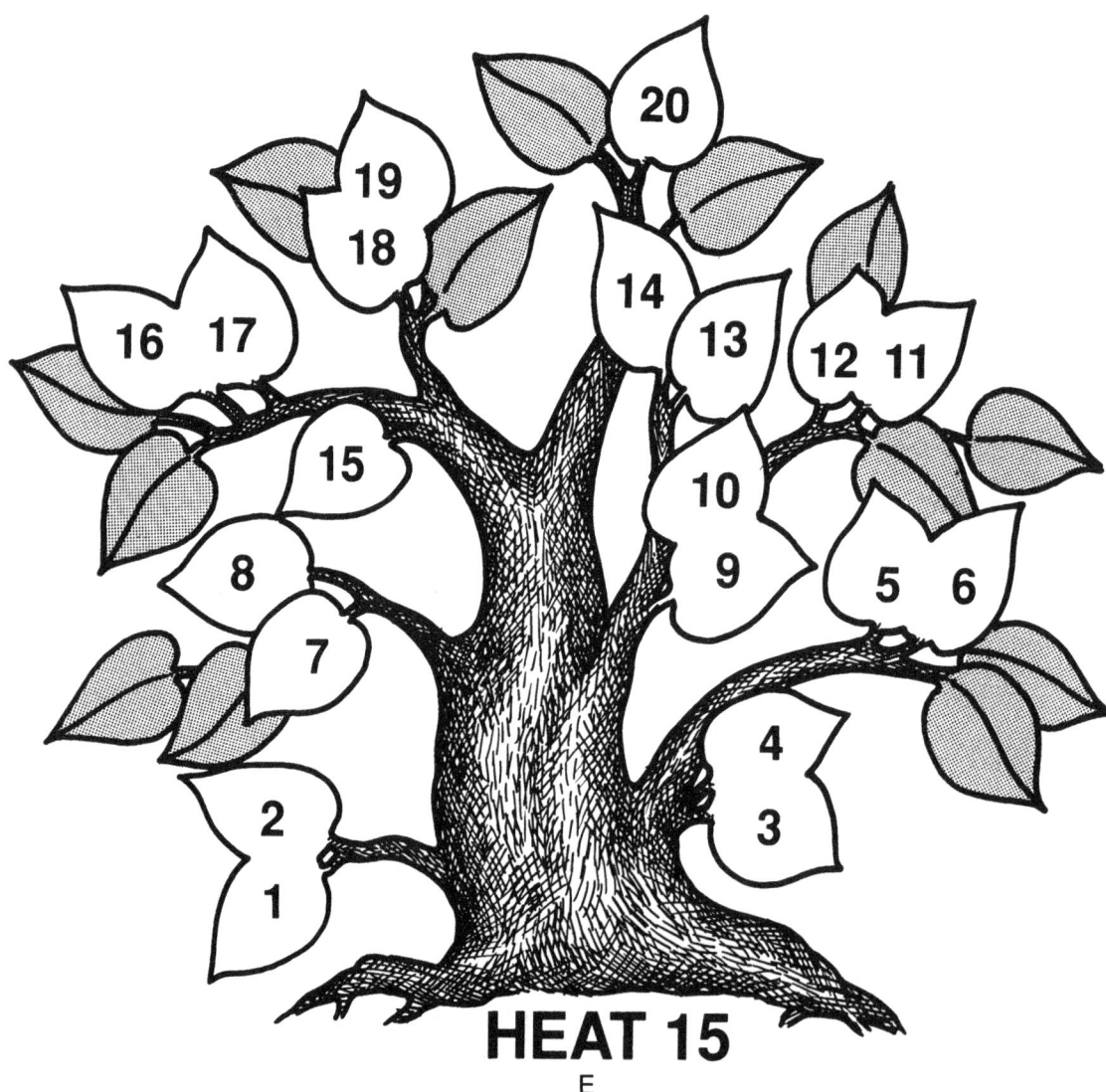

HEAT 15

E

LONG-RANGE OBJECTIVES

Given an environment rich in manipulatives. . .

Brains and muscles coordinate more smoothly as students interact with simple materials to improvise, engineer, construct and create.

PSYCHO-MOTOR

TOPS ACTIVITY

COGNITIVE

Students develop the full range of their intellectual capabilities. They learn to observe, question, test, analyze, predict, synthesize, evaluate and communicate.

Students will learn to learn. . . .

AFFECTIVE

An activity-centered environment helps learners succeed at their own levels. Students enjoy doing science because they feel positive about themselves.

Students will love to learn. . . .

Review / Test Questions

Photocopy the questions below. On a separate sheet of blank paper, cut and paste those boxes you want to use as test questions. Include questions of your own design, as well. Crowd all these questions onto a single page for students to answer on another paper, or leave space for student responses after each question, as you wish. Duplicate a class set and your custom-made test is ready to use. Use leftover questions as a review in preparation for the final exam.

task 1-2
Heat from a candle flame is conducted through a metal wire.
a. How do atoms in this wire interact to transfer heat energy?
b. What about the electrons?

task 1-2
"Conductron" is advertised as a new metal alloy that conducts heat better than copper. How could you test the validity of this advertising claim?

task 3
Why are metals better conductors of heat than nonmetals?

task 3
The metal spring on a clothespin feels cooler than the wood that surrounds it. Does this mean the spring has a lower temperature than the wood? Explain.

task 4
A goose down coat, a wool sweater and snow are all good heat insulators. What important property do they share?

task 5-6
Will a cup of hot chocolate stay hot longer if you put a lid on it? Why?

task 5-6
A space traveler brings along a candle and a book of matches, just in case the lights go out in her space ship. Will she have light in the event of a power failure? Explain.

task 7
To heat a pot of boiling water, is it more effective to apply the heat at the bottom of the pot or the top? Explain.

task 1-7
Can heat travel...
a. Through solids by convection? Explain.
b. Through liquids by conduction? Explain.

task 8
A piece of paper is wrapped tightly around a thick metal bar and held over a candle flame. Do you think the paper will catch fire?

task 9
Energy from the sun warms your face. Does it travel to your face by conduction, convection, or radiation? Explain your reasoning.

task 1-9
Explain how heat travels from the flame to points x, y and z near or on the metal bar.

task 10-12
Suppose you live in a hot desert climate.
a. What color would you paint your house? Explain.
b. What color would you paint your solar hot-water collector?

task 13
If carbon dioxide levels in the Earth's atmosphere continue to climb, what overall climate change could result? Explain.

task 14
The table lists cooling data for 2 different brands of thermos bottles. Graph this data to show which thermos is the most effective heat insulator.

time (min)	Therm-X (°C)	Vac-U (°C)
0	79.6	81.0
10	77.5	78.6
20	75.6	76.3
30	73.9	74.1
40	72.4	72.0
50	71.1	70.0
60	70.0	68.1

task 15-17
If you want to heat up a pot of water as fast as possible, would you fill it half full or all the way full? Explain.

task 15-17
A rock, heated by the sun, is thrown into the ocean. The temperature of the rock drops. Does the temperature of the ocean rise? Explain.

task 15-17
Exactly 200 ml of water at an initial temperature of 21.3° C is heated to 26.9° C. How many calories of heat did this water absorb?

task 18-19
If you take a bite of hot pizza, the tomato sauce is more likely to burn your mouth than the crust. Explain why.

task 18-19
Winds blowing across the ocean have less temperature variation than winds blowing across a desert. Explain.

task 20
A burning potato chip raises the temperature of 500 ml of water by about 6° C.
a. How many calories did the water absorb? How many food Calories is this?
b. Assuming the water captured 33 % of the total heat produced, how many food Calories are in this single chip?
c. How many chips would a child need to eat to fill his entire daily requirement of 2,700 Calories?

Answers

task 1-2
a. Atoms nearest the flame absorb energy by vibrating more energetically. These bump against adjacent atoms that, in turn, bump against other atoms in a chain reaction of increasing thermal energy.
b. The outer electrons in metals behave in a similar manner. They are not bound to any particular atom. Roaming freely through the wire, they speed up as they are heated, transmitting energy to other electrons through successive collisions.

task 1-2
Get 2 wires of equal diameter made from each kind of metal. Coat both with candle wax, then hold an end of each in a candle flame. Observe the pattern of melting wax as heat is conducted up both wires. If Conductron is a better conductor, its melting wax front will travel faster.

task 3
Metals are better conductors of heat because, unlike nonmetals, their outer electrons are free to flow through the metal. They absorb heat energy (kinetic energy) by moving faster, then transfer this energy to other electrons through successive collisions.

task 3
No. The clothespin is in thermal equilibrium with its surroundings. Both its metal and wood parts have the same temperature. The spring feels cooler because being metal, it conducts heat away from the hand faster than wood.

task 4
Goose down, wool and snow all trap significant amounts of air, which insulates each material against heat conduction.

task 5-6
Yes. The chocolate stays hot longer because the lid prevents heat from convecting out the top.

task 5-6
No. Less dense hot air will not rise above more dense cool air, because in the absence of gravity, everything is weightless. Without convection currents to lift oxygen-poor air above the flame and draw in fresh oxygen from the sides, the candle will quickly extinguish.

task 7
To heat a pot of boiling water, it is best to apply the heat at the bottom. This allows convection currents to carry hot water to the top, warming the pot uniformly. If the pot is heated only at the top, the hot water remains there because it is less dense. Heat can travel downward only by conduction, a very slow process.

task 1-7
a. Heat cannot travel through solids by convection because the atoms lack mobility. Being rigidly locked in place, they cannot expand and rise together in the manner of liquids.
b. Heat will travel through liquids by conduction if it is applied at the top of the liquid. Its atoms or molecules will vibrate with greater energy in response to this heat, and in the absence of upward convection, conduct energy very slowly downward as the atoms or molecules bump into their neighbors in a chain reaction.

task 8
The paper will not catch fire as long as the thick metal bar conducts sufficient heat away from it to keep the paper below its kindling point.

task 9
Heat travels from the sun to the face by radiation. It cannot travel through the intervening vacuum of space by conduction or convection because there is nothing in this vacuum to absorb energy. Heat travels this great distance with the speed of light as a form of electromagnetic radiation.

task 1-9
Heat travels from the flame...
    to point x by radiation.
    to point y by convection.
    to point z by conduction.

task 10-12
a. The house should be painted a light reflecting color inside and out. This will reduce heat transfer between the inside and outside to a minimum. On hot days, outside heat will be reflected away from the house. On cold nights, inside heat will be reflected back to its interior.
b. The tank should be painted black on the outside to absorb a maximum amount of radiated heat. It should be shiny on the inside (left unpainted if made of metal) to reflect internal heat back into the water.

task 13
Increasing levels of carbon dioxide in the atmosphere likely contribute to global warming. This is because carbon dioxide transmits higher-energy radiation coming from the sun, but absorbs lower-energy radiation that is radiated back from earth. Since the heat coming in exceeds the heat going out, temperatures gradually rise.

task 14

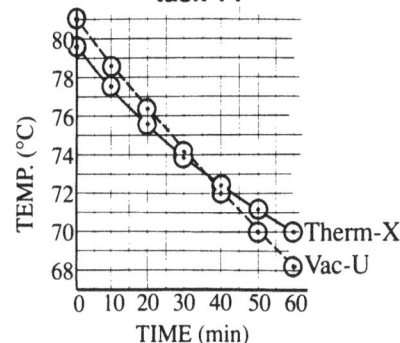

Therm-X is the best heat insulator.

task 15-17
Fill the pot only half full. Because it absorbs the same number of calories as the full pot, the temperature must rise twice as fast.

task 15-17
Yes. The heat energy lost by the rock must be gained by the ocean. But because the ocean is so vast, its rise in temperature is far too small to detect.

task 15-17
initial temperature = 21.3° C
final temperature = 26.9° C
difference = 5.6° C
200 ml water = 200 g
200 g x 5.6° C = 1,120 calories

task 18-19
The tomato sauce on the pizza contains more water than its crust. Because water has a very high specific heat, the sauce has a much higher heat capacity, and thus transfers relatively more heat to your mouth than the crust.

task 18-19
The desert (sand) has a much lower heat capacity than the ocean (water). It is thus prone to greater temperature extremes as it absorbs and radiates energy, heating and cooling the air above it much more rapidly than the ocean.

task 20
a. 500 ml water = 500 g
500 g x 6° C = 3,000 cal
$3,000 \text{ cal.} \times \dfrac{1 \text{ Cal.}}{1,000 \text{ cal.}} = 3 \text{ Cal.}$
b. 9 Calories.
c. $2,700 \text{ Cal.} \times \dfrac{1 \text{ chip}}{9 \text{ Cal.}} = 300 \text{ chips}$

TEACHING NOTES
For Activities 1-20

TASK OBJECTIVE (TO) trace heat flow through a wire. To understand how atoms in the wire interact to conduct heat.

HOT WIRE   O   **Heat (   )**

1. Cut a piece of iron wire about 10 cm long (as tall as this task card).

2. *Partly* melt a small lump of wax on a microscope slide. Roll the wire in this wax, just as it begins to resolidify, to thoroughly coat it with wax.

3. Hold the end of the wire just above the flame so it tilts slightly down. Describe how heat conducts (travels) from the flame through the wire.

WAX

4. Atoms in the wire vibrate more vigorously as they absorb heat energy. Propose a theory to explains how agitated atoms near the flame pass this energy along to atoms in the middle of the wire.

ATOM

FLAME

WIRE
(Magnified View)

5. Gases in a candle flame ignite at about 600° C. Wax liquefies around 60° C. Temperatures above 40° C feel uncomfortably hot. Describe the temperatures present in your hot wire.

© 1990 by TOPS Learning Systems   1

Answers / Notes

2. *Wax sticks to the wire only when it is cool enough to solidify. If students fully melt the wax, both the glass slide and wire will be too hot. No wax will stick until the temperature drops back down to the temperature of solidification.*

3. *The melted drops of wax should drip toward the flame. If the drips move up the wire, they carry heat up the wire as well.*

   The wax directly over the flame melts instantly, followed by wax adjacent to the flame, followed by wax further away, and so on, in a continuous melting front that advances towards the fingers. This front moves rapidly at first, then gradually slows down. It stops before reaching the fingers.

4. Atoms at the end of the wire that are exposed to the flame vibrate more energetically as they are heated. These atoms bump into adjacent atoms further down the wire that, in turn, bump into other atoms. Soon all the atoms are heated (vibrating vigorously). In this way, heat energy transfers through the wire without the atoms changing their positions.

5. The temperature of the wire drops from 600° C in the candle flame down to 60° C where the wax just melts. Beyond this point the temperature drops well below 40° C since it is still comfortable to hold.

Materials

☐ A roll of iron wire. Choose a diameter (perhaps 20 gauge) that is roughly the same thickness as aluminum wire and copper wire, also used in this module. A paper clip bent straight, or wire extracted from twist ties may be substituted. Avoid extremely thin wire.
☐ Wire cutters (optional). Or rapidly bend the wire back and forth until it breaks.
☐ A candle with drip catcher plus matches.
☐ A microscope slide.

(TO) discover that some metals conduct heat more rapidly than others. To understand how electrons interact in wire to conduct heat.

HEAT RACE ○ Heat ( )

1. Coat 3 pieces of wire with wax as before. Each should be about 10 cm long, the height of this card.

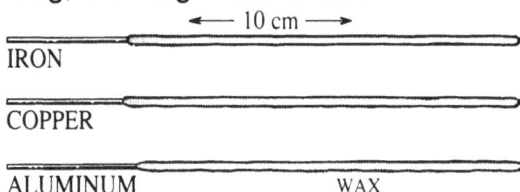

← 10 cm →

IRON ▬▬▬▬▬▬▬▬▬▬

COPPER ▬▬▬▬▬▬▬▬▬▬

ALUMINUM ▬▬▬▬▬▬▬▬ WAX

2. Hold heat races: Place 2 wires side by side, tipped slightly down, over a candle flame. Summarize your results in a table.

Race	Result
iron vs. copper	
iron vs. aluminum	
copper vs. aluminum	

3. Heat is conducted through metals because atoms vibrate against each other. Even more importantly, their free outer electrons also collide with each other, and other atoms.

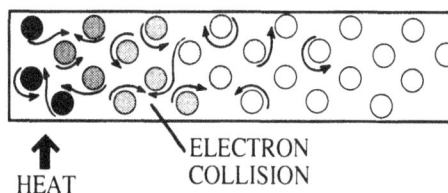

HEAT ↑ ELECTRON COLLISION

   a. Which metal likely holds its outer electrons most tightly? What makes you think so?

   b. Would you expect good conductors of heat to be good conductors of electricity (electrons) as well? Explain.

2

Answers / Notes

2.

Race	Result
iron vs. copper	Copper easily wins race.
iron vs. aluminum	Aluminum easily wins race.
copper vs. aluminum	Both are very fast. Copper usually wins.

3a. Iron likely holds its outer electrons more tightly because they don't transfer energy as rapidly as the outer electrons of aluminum and copper atoms.

3b. Yes. Since heat energy and electric energy are both transmitted by moving electrons, the easier these electrons flow, the better each form of energy travels through its wire.

Materials

☐ A roll of aluminum wire and a roll of copper wire (perhaps 20 gauge) that is roughly the same thickness as the iron wire used in the first activity.
☐ Wire cutters (optional). Or rapidly bend the wire back and forth.
☐ A candle with drip catcher plus matches.
☐ The glass microscope slide used previously to melt wax.

(TO) contrast the heat conduction properties of a conductor with an insulator.

CONDUCTORS / INSULATORS ○   Heat (   )

GLASS   ALUMINUM

1. Cut a strip of aluminum from the bottom of a pie plate, equal in size to a microscope slide.

2. Coat 1 surface of each with melted wax, and let them cool.

WAX

3. Hold another heat race: place a corner of both "slides" side by side over a candle flame. Write your observations in words and pictures.

4. Substances that slow the passage of heat are called insulators. Which is the insulator and which is the conductor?

5. Wax that gets very hot vaporizes to a gas. Did any parts of either slide get this hot? Why?

6. Metals have free outer electrons. Do you think this is true for nonmetals like glass?

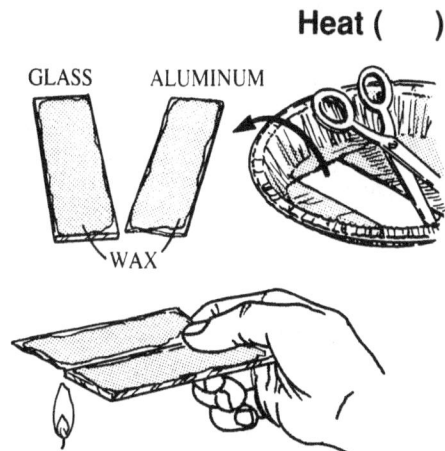

© 1990 by TOPS Learning Systems   3

Answers / Notes

3. There is a dramatic difference between glass and aluminum. While the candle heat spreads through only a small corner of the glass slide, it completely melts all the wax on the aluminum rectangle.

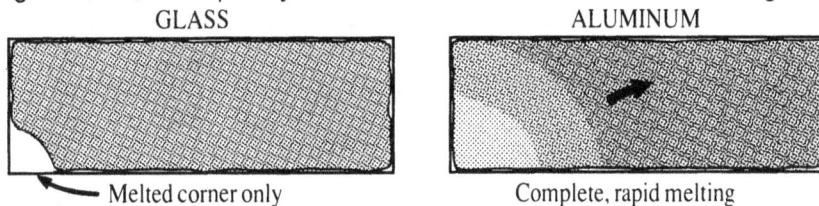

GLASS   ALUMINUM

Melted corner only   Complete, rapid melting

4. Glass is the insulator, while aluminum is the conductor.

5. Yes. The corner of the glass slide directly under the candle flame became hot enough to vaporize its wax, leaving a clean, wax-free surface. Acting as a heat insulator, it confined the energy to a small area directly over the flame, allowing temperatures to increase enormously. The aluminum slide, by contrast, quickly conducted heat to other parts of the slide before temperatures could reach the vaporization level.

6. No. The outer electrons in glass and other nonmetals are tightly bound to their respective atoms. These electrons cannot travel between atoms and collide with one another as they do in metals. As a result, nonmetals are poor conductors of heat.

Materials

☐ Scissors.
☐ A recyclable aluminum pie plate. These are relatively thin, and can be cut with ordinary scissors.
☐ A candle with drip catcher plus matches.
☐ The glass microscope slide previously used to melt wax.

(TO) observe that heat flows from higher-temperature objects to lower-temperature objects. To recognize that air is a very good insulator.

COLD FINGER　　　　　　　　○　　　　　　　　Heat ( )

1. Set an ice cube in a pie tin and touch it with your finger.

   a. Explain why your finger begins to feel cold. (Hint: Heat travels, not cold. Cold is merely the absence of heat energy.)

   b. How long could heat continue to flow in this experiment? Think carefully!

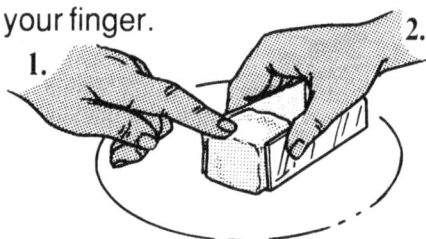

2. Pinch the ice cube between a glass microscope slide and the aluminum slide you made previously. Does one side feel colder than the other? Explain.

3. Puff up a plastic sandwich bag somewhat, then seal it air-tight with a strip of masking tape.

4. Wrap your ice cube in the baggy, cushioned by air.

   a. Evaluate air as an insulating material.

   b. Compare the insulating properties of air with plastic.

5. Do you think a vacuum (nothing at all) would be a good insulator? Why do you think so?

© 1990 by TOPS Learning Systems

4

Answers / Notes

1a. The finger feels cold because it loses heat to the ice cube. Heat flows out of the finger and into the ice.

1b. Heat continues to flow as long as there is a temperature difference between your finger and what you touch. The ice will eventually melt and the water will reach room temperature. But even then heat still flows from your finger because it is warmer than what you touch: first ice, then water, then the pan (after the water evaporates). As long as you remain alive, heat flows from your warmer finger into the colder environment.

2. Yes. The aluminum slide feels colder than the glass slide. This is because heat from your hand travels faster through aluminum (which is a conductor) than through glass (which is an insulator).

4a. Air is an excellent insulator. Even the smallest air gap between the fingers and the ice cube blocks nearly all heat loss to the ice. The ice does not feel cold at these points.

4b. Air is a much better insulator than plastic. Where skin, plastic and ice come into direct contact, with no intervening layer of air in between, the ice continues to absorb heat and feel cold.

5. Yes. Heat will not conduct through a vacuum because there are no atoms or molecules to transfer the energy. *(Heat will travel through a vacuum, in a manner similar to light, as radiant energy. This concept is introduced in activity 10.)*

Materials

☐ An ice cube.
☐ An aluminum pie tin. A plate will also serve.
☐ The glass microscope slide and aluminum "slide" from activity 3.
☐ A plastic sandwich bag.
☐ Masking tape.

(TO) observe that hot water is lighter (less dense) than cold water. To understand why heated fluids convect upward.

RISE AND FALL ○ Heat ( )

1. Mix 1 drop of blue food coloring in a jar of cold water. Fill a smaller jar brim full, using a pie plate to catch any overflow.

COOL BLUE WATER

2. Fill a second jar with hot water, brim full like the first. Keep both in the same overflow container.

CLEAR WARM WATER

3. Cover the cool blue water with an index card, and carefully invert it over the warm clear water. Carefully pull out the card and write your observations.

REMOVE CARD

4. Repeat the experiment. This time invert the hot clear water over the cool blue water. Write your observations.

5. Warm water rises as cool water sinks. Heat movement of this kind is called *convection*.
   a. Did you observe convection in steps 3 and 4? Explain.
   b. Do atoms *conduct* heat through a wire in the same manner as molecules *convect* heat through a fluid? Explain.

© 1990 by TOPS Learning Systems   5

Introduction

Atoms and molecules in a fluid (a liquid or gas) move more energetically as they absorb heat. As a result the same number of particles occupy more space. This makes the fluid lighter (less dense), causing it to rise above the cooler, heavier (more dense) fluid that takes its place. This process is called convection.

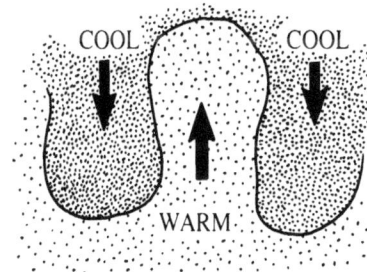

COOL   COOL

WARM

Answers / Notes

3. After removing the card, the blue water sinks while the clear water rises, rapidly mixing the 2 liquids.

4. After removing the card, the clear hot water remains in place above the cool blue water below. There is no mixing.

5a. Convection was observed in step 3 only: the lighter clear warm water moved upward, displacing the heavier cool blue water that sank below. There was no convection in step 4 because the lighter clear warm water was originally placed on top. It could rise no higher.

5b. No. Heat conducts through a wire because electrons collide and atoms vibrate. The atoms themselves stay in one place. Heat convects, by contrast, as warmer parts of a fluid rise above colder parts. Molecules in the fluid actually move higher and lower in great numbers.

Materials

☐ Blue food coloring dispensed in a dropper bottle.
☐ Warm and cold water. Use it straight from the tap if available. Temperature differences need not be extreme.
☐ Two larger glass jars for pouring water plus 2 smaller jars to contain it. Pint jars and baby food jars are suitable. The smaller jars must have round, level mouths. Beakers with pour spouts won't work.
☐ A pie plate or equivalent to contain water.
☐ An index card.

(TO) build a simple windmill that will rotate in the convection currents generated by a burning candle.

CONVECTION MACHINE ◯ **Heat ( )**

1. Cut out a coil from an index card.

2. Tape a thread to the center of your coil, then suspend it at least a hand span above a burning candle. Be careful it doesn't catch fire.

3. Tell how your convection machine works. Illustrate your answer with a diagram.

4. What makes heated air rise?

6

Answers / Notes

2. *The coil should be held well above the flame, where the updraft is broad enough to surround it entirely. It will not turn well near the flame, nor is it safe to hold it so close.*

3. The candle flame creates an upward convection current that pushes the spiral coil at an angle, causing it to turn.

4. As the air is heated, its atoms and molecules move faster, and thus take up more space. This expansion makes the air less dense, causing it to rise above adjacent cooler air of higher density that sinks underneath to displace it.

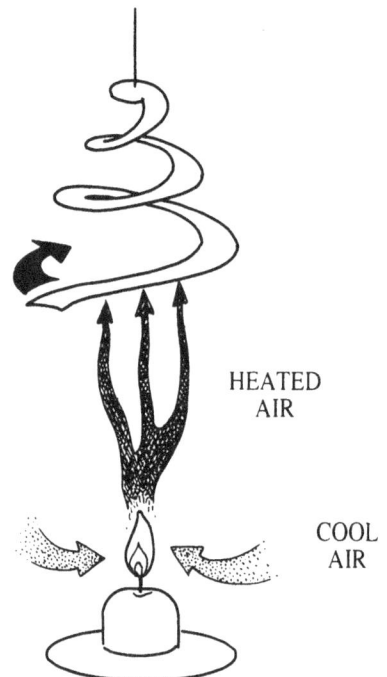

HEATED AIR

COOL AIR

Materials

☐ Scissors.
☐ An index card. Scratch paper may be substituted, but the coils will stretch out to a greater degree.
☐ Thread.
☐ Tape.
☐ A candle with drip catcher, alcohol lamp or Bunsen burner, plus matches.

(TO) appreciate that water is a good convector of heat, but a poor conductor.

TOO HOT TO HANDLE?　　　O　　　　　　　Heat (　)

1. Fill a test tube 1/5 full of crushed ice. Push a small plug of steel wool over the ice with your pencil, then fill the tube with water.

2. Try to boil water at the top of the tube without melting the ice at the bottom. Is water a conductor of heat, or an insulator? Explain.

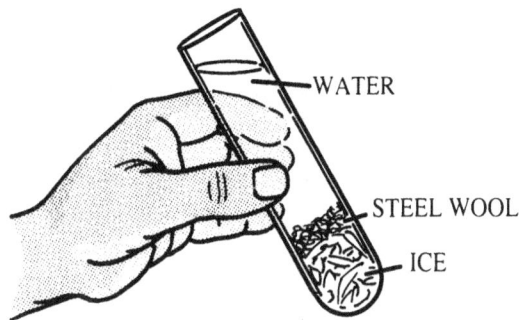

WATER

STEEL WOOL

ICE

3. Would this experiment turn out the same with ice at the top of the tube and the flame at the bottom? Why?

7

Answers / Notes

2. Water is an insulator. Heat energy that was absorbed by the water at the top of the tube was not conducted downward to any significant degree. It remained localized and undissipated at the top of the water, causing its surface to boil. The water directly below remained cool, and the ice at the bottom stayed frozen.

3. No. Heating the water at the bottom of the tube would set up convection currents. Hot water would expand and rise to the top of the test tube, carrying enough heat energy to melt the ice. Cool water would sink below where it could be warmed. Mixed in this manner, all parts of the water would uniformly approach boiling.

Materials

☐ A test tube. A small size works best so the water can be brought to a quick boil.
☐ A candle with drip catcher, alcohol lamp or Bunsen burner, plus matches.
☐ Matches.
☐ Crushed ice. Ice cubes may be crushed by wrapping them in a towel and pounding them with a large jar or other heavy object.
☐ Steel wool.

(TO) discover that paper in direct contact with a candle flame won't burn, as long as its heat is conducted away.

PAPER COOKING POT   O   Heat (   )

1. Press a scrap of notebook paper into the mouth of a test tube with your thumb.

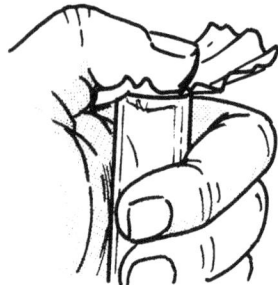

2. Cut it into a spoon shape. Slip a paper clip over the handle so its end supports the bowl.

3. Trim off any excess paper, then attach a clothespin. Fill with water and heat it over a candle flame. Write your observations.

SUPPORT THE BOWL

4. Water boils at 100° C (at sea level). It will not get hotter than this until all the liquid is vaporized to steam.

    a. Is the combustion temperature of paper higher than 100° C? Explain.

    b. Will the spoon eventually burn if you leave it over the flame? Explain.

5. Explain the role of conduction and convection in keeping the paper cool.

© 1990 by TOPS Learning Systems   8

Answers / Notes

3. The water boiled but the spoon did not catch fire. *(Students should take care not to expose the paper spoon handle to the candle flame. It will catch fire, because it is not in direct contact with water, which conducts and convects heat away.)*

4a. Yes. The paper could not reach its combustion temperature and ignite as long as water remained in the spoon.

4b. Yes. After the water boils and evaporates away, it will no longer remove heat from the paper. The temperature of the paper will rise to its combustion temperature, and it will ignite.

5. Heat from the candle flame is conducted through the paper to the water. It warms the water, which convects to the surface and evaporates as steam.

Materials

☐ Notebook paper.
☐ A test tube.
☐ Scissors.
☐ A paper clip.
☐ A clothespin.
☐ A candle with drip catcher, alcohol lamp or Bunsen burner, plus matches.

understand that heat energy also travels like light, as an electromagnetic wave. To distinguish between conduction, convection and radiation.

RADIATION   O   Heat ( )

1. Probe the edge of a candle flame with wax shavings stuck to the end of a toothpick. Sketch the *isotherm* (an equal-temperature line) surrounding the flame where this wax just starts to melt (about 60° C).

2. Suppose candle wax just melts at point A, part of your 60° C isotherm.
    a. Heat cannot travel to point A by convection. Why not?
    b. Heat cannot travel to point A by conduction. Why not?

3. Heat travels to point A by *radiation*.
    a. Name our most important source of radiated heat.
    b. How fast do you think radiated heat travels?

4. Thoroughly heat an iron wire in the hot flame of an alcohol lamp or Bunsen burner. What evidence can you see that radiated heat energy is similar to light?

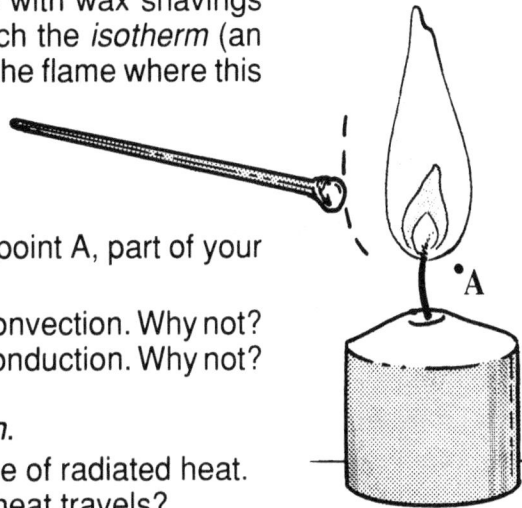

9

Introduction

Radiant heat energy travels through space as a wave, partly electric and partly magnetic. It fits between light and microwaves on the electromagnetic spectrum:

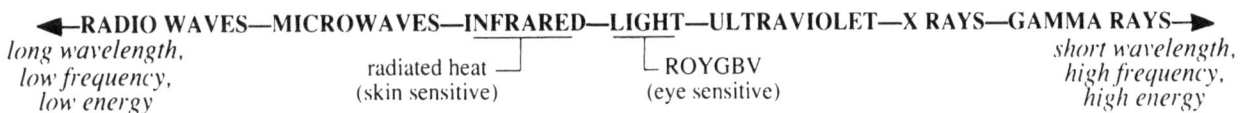

◄—RADIO WAVES—MICROWAVES—INFRARED—LIGHT—ULTRAVIOLET—X RAYS—GAMMA RAYS—►

long wavelength,   radiated heat ⌐   ∟ROYGBV   *short wavelength,*
low frequency,   (skin sensitive)   (eye sensitive)   *high frequency,*
low energy   *high energy*

Answers / Notes

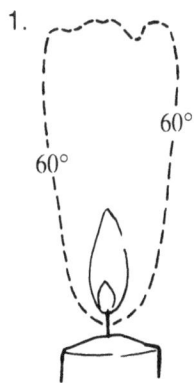

1.

2a. Convected heat can only rise. Yet no part of the flame is directly below point A.

2b. There is nothing between point A and the flame except air, which has already been shown to be a very good insulator. Moreover, this insulating air doesn't remain long enough to be heated up. It is continually being drawn into the flame to replace rising convected air.

3a. The sun.

3b. The sun's energy comes to earth through deep space over great distances. When its light strikes the eyes, its heat warms the face. This suggests that radiated heat and light travel together at the speed of light.

4. As the wire heats up it begins to glow, first dull red and finally bright orange. This suggests that radiated heat is just below light on the electromagnetic spectrum in the infrared region. As the wire warms to higher energy levels some of its energy reaches the wavelength of visible light.

Materials

☐ A candle plus drip catcher and matches.
☐ A toothpick.
☐ Iron wire. Copper and aluminum wire conduct heat too well to achieve temperatures in the visible spectrum (about 500° C). Straightened bobby pins work great, after you peel or melt off the plastic tip. The thin iron wire in twist ties is also suitable.
☐ A hot flame source. Use an alcohol lamp or a Bunsen burner. Candle flames are too cool.

(TO) observe that radiated heat energy tends to be absorbed by a dull black surface, and reflected by a shiny metallic or white surface.

REFLECTION/ABSORPTION ◯   Heat (   )

1. Clamp clothespins onto 2 tin can lids. Cover just 1 side of 1 lid with candle soot; leave the other lid shiny on both sides.

2. Drip melted candle wax in the middle of each lid (*not* on the blackened side). Quickly stick a penny to each lid before the wax dries.

SHINY
SOOTY

PENNY

3. Hold each lid so the pennies are directly opposite the candle flame, behind each lid. Keep them a small but equal distance away.

   a. Which penny drops first?
   b. Explain your observations in terms of absorption and reflection.

4. Light is absorbed and reflected like radiated heat. Explain why this makes sense.

SHINY   SOOTY

© 1990 by TOPS Learning Systems   10

Answers / Notes

3a. The penny on the blackened lid drops first.

3b. The shiny surface tended to reflect heat radiating from the candle flame, whereas the black sooty surface more readily absorbed it. Hence, the shiny lid stayed relatively cool, while the sooty lid captured enough heat to melt the wax and drop the penny.

4. The unsooted lid appears shiny because light (like radiated heat) reflects off its surface into the eye. But the sooted lid appears dull and black because most of the light that strikes its surface (like radiated heat) is absorbed.

Materials

☐ Two tin can lids of similar size and appearance. Aluminum or bronze colored surfaces are equally suitable as long as they are shiny.
☐ Two clothespins.
☐ A candle with drip catcher plus matches.
☐ Two pennies.

(TO) test black, white and shiny metallic surfaces as emitters of heat radiation. To distinguish between the 3 processes of heat transfer.

BEST EMITTER? ○ Heat ( )

1. Tightly wrap a clean, dry test tube in each material. Use just enough to fully cover the glass. Hold the wrapper with clear tape.

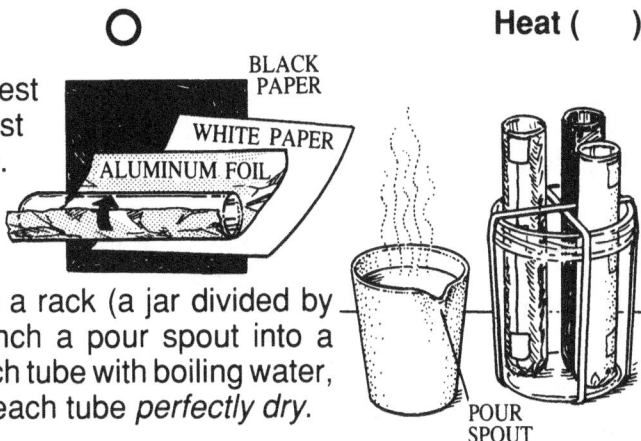

BLACK PAPER
WHITE PAPER
ALUMINUM FOIL

POUR SPOUT

2. Stand the tubes upright in a rack (a jar divided by rubber bands will serve). Pinch a pour spout into a styrofoam cup. Use it to fill each tube with boiling water, while keeping the outside of each tube *perfectly dry*.

3. Wait 10 minutes or longer for the test tubes to cool. While you are waiting…
   a. *Briefly* touch the side of each tube. Which feels hottest? Why?
   b. Predict which tube will be the best heat emitter (will cool the fastest).

4. After 10 minutes or so, use a thermometer to quickly measure the temperature of the water in each tube. Do this in the same order you filled them.

5. Was your prediction correct? Propose a theory to explain your observations. (Seal your filled test tubes with corks or clay. Save them for the next activity.)

© 1990 by TOPS Learning Systems   11

Introduction
Discuss the various ways that heat is lost to the environment by a test tube of steaming hot water: by conduction into the table surface; by upward convection through water and air; by radiation in all directions.

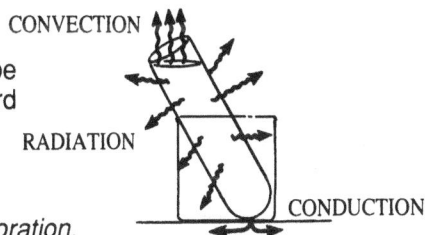

CONVECTION
RADIATION
CONDUCTION

Answers / Notes
2. *Water spilled on the outside of the test tubes will cool them by evaporation.*

3a. The tube covered with aluminum foil feels hottest because aluminum is the best conductor of heat. *(Good observers may report that the black paper feels slightly warmer than the white paper.)*

3b. Let all predictions stand. *(Because the foil feels hottest, many will incorrectly predict that it is the best emitter.)*

4. *Students should take temperature readings as rapidly as possible, in the same order as the hot water was originally added. Only the bulb should contact the water, not the whole thermometer. Keep it submerged until the mercury stops rising (perhaps 5-10 seconds).*

Tube temperatures should vary by 1 to 2 °C.

— BLACK ——— WHITE ——— SHINY →
(coolest)   (warmest)
best emitter   worst emitter

5. Aluminum best conducted heat into the fingers when the test tubes were touched in step 3. But conduction does not account for much heat loss, since each test tube is surrounded by insulating air. Heat loss through the top of each tube by convection is more significant. Yet convection fails to explain why the tubes cooled at different rates, since they all had openings of equal diameter. Only heat loss by radiation fully explains the results. The shiny foil surface facing inside the tube best reflected radiant heat back into the water, keeping it warmest; white paper also reflected energy, though not as well; the black paper surface best absorbed energy and radiated it to the outside, cooling the water most rapidly.

Materials
☐ Three test tubes.
☐ Aluminum foil.
☐ Clear tape.
☐ Scissors.
☐ A thermometer.

☐ Black paper and white paper. These should have roughly the same thickness. Ordinary copy paper will work for white. To get black, students might cut a white rectangle of suitable size, then color both sides with a marking pen or crayon.
☐ A test tube rack. Or divide a jar into quarters with rubber bands.
☐ A source of boiling hot water. Use a Bunsen burner or hot plate plus teapot.
☐ Corks or lumps of clay to seal the test tubes.

(TO) test black, white and shiny metallic surfaces as absorbers of heat radiation.

BEST ABSORBER?   O   Heat (   )

1. Get your stoppered test tubes full of water from the previous activity. They should all be at room temperature.

2. Expose these test tubes to about 10 minutes of direct sunlight outdoors, or use a heat lamp or high wattage light bulb indoors. Slant the tubes to receive perpendicular rays.

3. While you are waiting, predict which tube should best absorb heat (warm the fastest). Recall results from previous activities to support your answer.

4. After about 10 minutes, quickly measure the final temperature of the water in each tube, while shielding all tubes from additional radiation.
    a. Evaluate your prediction. Was the best absorber also the best emitter?
    b. What would happen if a good absorber were a poor emitter?
5. When does a test tube emit heat radiation, and when does it absorb it?
6. What color should you paint your house to conserve energy? Explain.

© 1990 by TOPS Learning Systems   12

Answers / Notes

3. The test tube covered by black paper should absorb the most radiated heat. In activity 10, the blackened lid best absorbed radiated heat, while the shiny metal surface best reflected it. Further, in activity 11, the black-papered test tube again best absorbed radiated heat from the hot water inside and radiated it to the outside, while the white paper and aluminum foil tended to reflect this heat back into the water.

4. Again, the temperatures in each tube should vary by 1 to 2 °C.

$$\text{— SHINY ——— WHITE ——— BLACK} \rightarrow$$
    (coolest)   (warmest)
  worst absorber   best absorber

4a. As predicted, black was the best absorber. Black was also the best emitter.
4b. It would readily warm up but not cool down, and thus remain warmer than other objects nearby.

5. The test tube emits *heat* when it is *warmer* than its surroundings, and *absorbs* heat when it is *cooler* than its surroundings.

6. Paint your house white. (Or if you can stand the color, paint it a shiny, metallic aluminum.) White and shiny aluminum are poor emitters: in winter the heat inside your house tends to stay inside. White and shiny aluminum are also poor absorbers: in summer the heat outside your house tends to stay outside.

Materials

☐ The three stoppered test tubes from activity 11. One is covered with shiny foil, one with white paper and one with black paper.
☐ Something to prop the test tubes against. A brick, a rock, or books will serve.
☐ A thermometer.
☐ Direct sunlight. Indoors, expose test tubes to a heat lamp or high wattage bulb.

(TO) construct a functioning greenhouse. To use it as a model for explaining how carbon dioxide gas in our atmosphere acts to warm the Earth.

THE GREENHOUSE EFFECT O   Heat (   )

Fact A: The sun (any hot body) radiates higher-energy short waves. The earth (any cool body) radiates lower-energy long waves.

HIGH ENERGY
LOW ENERGY
EARTH

Fact B: Glass transmits shorter light waves, but not longer infra-red waves.

LIGHT
HEAT

1. Fill 2 small jars about 1/6 full of sand and just cover with water. Cap one with a third jar and seal with masking tape. Leave the other uncovered.
   a. Will the wet sand in each jar warm equally if you set both jars in the sun? Apply facts **A** and **B** as you answer.
   b. Test your prediction for 10 minutes (or longer) and report your results.

TAPE
SAND & WATER

2. Carbon dioxide in Earth's atmosphere is transparent to shorter wave lengths and absorbs longer wavelengths, in a manner similar to glass.
   a. Explain why it is important to reduce our use of gasoline, coal and oil.
   b. What can *you* do to make a difference?

EARTH

13

Answers / Notes

1a. No. Higher-energy, shorter-wavelength radiation (light) passes through glass directly to the wet sand in both the covered and uncovered jars. This heats the wet sand which, in turn, radiates lower-energy, longer-wavelength energy that is reflected by the greenhouse glass. This reflected radiation is trapped by the closed greenhouse but escapes through the top of the open one. *(Significant heat is also lost through the open top by convection.)*

1b. In direct sunlight the temperature of the wet sand in the closed greenhouse rises faster, producing a difference of about 1° C every 10 minutes over the first half hour.

2a. These fossil fuels emit carbon dioxide when we burn them for energy

$$fuel + O_2 \text{ ----- } energy + CO_2 + H_2O$$

Carbon dioxide, like greenhouse glass, transmits short wavelength heat radiation from the sun, but absorbs longer wavelength heat radiation emitted by the warmed earth. With more energy coming to the earth and less energy radiated back into space, our globe is gradually getting warmer. *(Over the last half million years temperatures on earth have averaged between 19 °C and 27 °C, causing ice ages to come and go. Presently we are at the high end of this range, and warming.)*

2b. Use energy wisely. Conserve.

Materials

☐ Three baby food jars.
☐ Sand. Keep the quantity in each jar to a minimum, perhaps 20 ml per jar. This will cause temperatures in the wet sand to rise more rapidly. (Sand is added to provide a more realistic earth model in step 2. Water alone can be substituted without affecting the results.)
☐ Masking tape.
☐ Direct sunlight. Indoors, expose model greenhouses to a heat lamp or high wattage bulb.
☐ A thermometer.

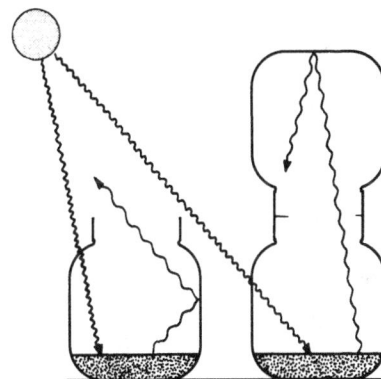

(TO) compare cooling curves for glass and metal containers.

HOLD THAT HEAT   O   Heat ( )

1. Mark where 100 mL of water reaches inside a styrofoam cup.

2. Fill it to this mark with near-boiling water, then pour it into a metal can. Record the highest temperature reached while you note the position of a second hand on your watch or a wall clock.

3. Mark the temperature of the water after each new minute for the next 10 minutes. Stir the water once or twice before reading each temperature.

4. Repeat this experiment for a glass jar. Present your time and temperature data in a table, and graph your results.

5. Interpret your data.

6. Which forms of heat loss made the most difference in this experiment? Explain why you think so.

time (min)	temperature (°C)	
	metal can	glass jar
0		
1		
2		
3		
⋮		
10		

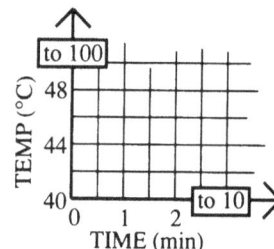

© 1990 by TOPS Learning Systems

14

Answers / Notes

2-4.

time (min)	temperature (°C)	
	metal can	glass jar
0	88.2	84.8
1	83.0	80.0
2	78.5	76.1
3	74.0	72.8
4	70.6	69.8
5	67.8	67.1
6	65.0	65.0
7	62.9	63.0
8	60.3	61.1
9	58.6	59.4
10	56.9	58.0

5. The glass jar held heat slightly better than the metal can: its cooling curve was a little shallower. Both containers had curved graph lines, suggesting faster heat loss at higher temperatures and slower heat loss at lower temperatures.

6. Heat loss through conduction likely caused most of the temperature difference between the glass jar and the metal can. Free electrons in the metal conducted more heat energy into the table, while the glass container better insulated against this transfer. Both containers lost roughly equal amounts of heat by convection, since their open tops have roughly equal area. And both the glass and metal containers trap radiant heat energy, reflecting it back into the water.

Materials

☐ A styrofoam cup.

☐ A 100 mL graduated cylinder.

☐ A metal can and a glass jar of roughly equal size. In general, the smaller the container and the smaller its opening, the better it will hold heat. The results above were obtained with a 15 ounce "tin" can and a pint jar.

☐ A source of near-boiling water. In this activity and others, it is most convenient and economical to heat a kettle of water on an electric hot plate that students can use as needed. With Bunsen burners, of course, students can heat their own.

☐ Thermometers.

☐ Graph paper. Photocopy the grid at the back of this book.

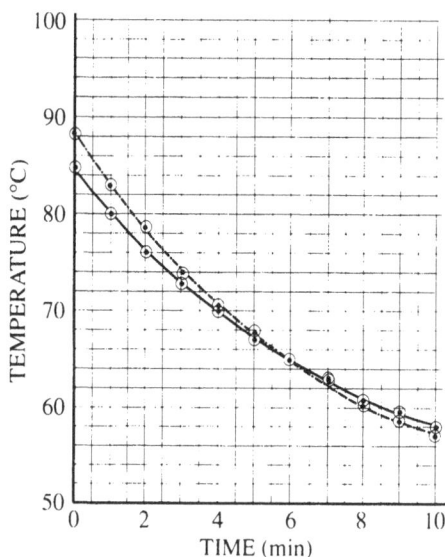

Extension

Compare the cooling curves for 2 identical containers, one that is covered with a lid and one that is left open.

(TO) define and use calories as a unit of heat energy. To distinguish between temperature and heat.

COLD HANDS   O   Heat (   )

1. Add 1000 g of room-temperature water to a large container. Accurately record its temperature to the nearest .1° C.

2. Completely submerge your *right* hand in this water for exactly 5 minutes. (Count each minute so you don't lose track.) Accurately record the final temperature and calculate the increase.

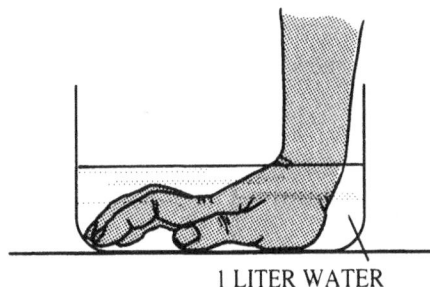
1 LITER WATER

   a. One *calorie* (a unit of energy) raises 1 gram of water 1° C. How many calories of heat are required to raise the water in your container 1° C?
   b. How many calories did your hand transfer to the water?

3. Assume your *left* hand gives off the same amount of heat as your right.
   a. Predict how much your left hand will raise the temperature of 500 g of water in 5 minutes. Show your calculations.
   b. Test your prediction. (Keep your *whole* hand under water as before.)

4. Does a thermometer measure temperature or heat? Refer to your results in steps 2 and 3 as you answer.

15

Answers / Notes

1. *The density of water is 1 g/mL. 1000 g thus occupies 1000 mL, or 1 liter.*
2. Here is one possible result for the right hand:
    final temperature = 21.7° C
    initial temperature = 20.4° C
    difference = 1.3° C
2a. If 1 calorie raises 1 g of water 1° C, then 1,000 calories raises 1000 g of water by the same 1° C.
2b. 1000 g x 1.3° C = 1,300 calories.

3a. Half as much water should be raised to twice the temperature given the same amount of heat. Thus, 500 g x 2(1.3° C) = 1,300 calories temperature difference = 2.6° C.
3b. Agreement between hands is very close, considering measuring uncertainty in the thermometer.
    final temperature = 23.0° C
    initial temperature = 20.5° C
    difference = 2.5° C

4. A thermometer measures temperature *(the average kinetic energy of the water molecules)* but not heat *(the total kinetic energy of the water molecules).* It recorded, for example, that the 500 g water sample warmed twice as much as the 1000 g sample, even though both received the same amount of hand heat.

Discussion

Food Calories are really kilocalories. To distinguish them from smaller calorie units, they are written with a capital C: 1000 calories = 1 Calorie. An adult gives off about 2,400 Calories of heat per day. It is necessary to consume that much food on a daily basis to maintain body weight.

a. How many Calories per hour does an average adult conduct, convect and radiate as heat? (100)
b. How many calories per hour? (100,000)
c. How many calories per minute? (1,667)
d. How many calories per minute were lost by *both* hands in room-temperature water? (2,600 / 5 = 520)

e. What can you conclude? (Almost 1/3 of an adult's average heat loss per minute passed through the hands when placed in room temperature water. Since our hands make up far less than 1/3 of our total body surface, this shows that water is a much better conductor of heat than air. An excellent way to lose weight would be to take up swimming in an unheated pool!)

Materials

☐ A large (gallon) plastic milk container cut in half, or equivalent. If you are substituting other containers, make sure they are large enough to accommodate your hand plus a full liter of water, yet small enough to completely submerge your hand when only a half liter of water is added.
☐ A large graduated cylinder. 100 mL graduates are OK, but it takes considerable time to fill the jug. As an alternative, supply liter or quart jars with the 500 mL and 1000 mL levels clearly marked.
☐ Room-temperature water and a thermometer.
☐ A calculator (optional).

(TO) discover that the amount of heat given off by hot water equals the amount of heat absorbed by cold water.

WATER MIX (1) ○ Heat ( )

1. Use a graduated cylinder to accurately calibrate the inside of 2 styrofoam cups. Mark the 50 mL, 75 mL and 100 mL levels.

2. Measure 75 mL of cool water into one cup and record its temperature. Do the same for 75 mL of hot water. Quickly pour the cool into the hot and immediately record the final temperature.

75 mL COOL

75 mL HOT

3. Calculate the change in temperature for each water sample.

4. Compute the number of calories gained by the cool water and lost by the hot water.

5. Within the limits of experimental error, does the heat gained by the cool water equal the heat lost by the hot water? What are some sources of error?

(Save your graduated cups.)

16

Answers / Notes

2-4. *Notice how taking the temperature of the hot water is delayed until just before the water samples are mixed. Moreover, the cool water is poured into the hot, not the other way around. These measures minimizes heat loss to the surroundings. This is especially important if you are using water that is near boiling.*

Caution students to avoid leaving heavy glass thermometers unattended in light styrofoam cups. They are not very stable, and could be easily knocked over and broken.

Sample data:

2. temperature of cool water = 24.0° C
   temperature of hot water = 52.1° C
   final temperature of mix = 37.9° C

3. Change in cool water:
   37.9° C - 24.0° C = 13.9° C
   Change in hot water:
   52.1° C - 37.9° C = 14.2° C

4. Heat gained by cool water:
   (75 g) (13.9° C) = 1043 calories
   Heat gained by hot water:
   (75 g) (14.2° C) = 1065 calories

5. Yes. Heat gained = Heat lost. Sources of error in this experiment include:

•heat exchange: Heat is continually being lost to the cup and its surroundings by the hot water and gained from the cup and its surroundings by the cool water.

•water quantity: The calibrations on the side of the cup are only a rough measure of volume. Possibly a little more or a little less than 75 mL of water were used for each sample.

•temperature readings: You must allow the bulb to reach thermal equilibrium before recording the temperature. But you must not wait so long that movement of the mercury is caused by heat exchange with the surroundings. When you do read the thermometer, there is considerable measuring uncertainty in the tenth's place.

Materials

☐ Styrofoam cups. A standard coffee-cup size has the capacity to accommodate the required 150 mL of water.
☐ A 100 mL graduated cylinder.
☐ Cool and hot water. Take this as it comes from the tap, if available. Otherwise, keep heated water in a teapot or ask students to warm their own using a Bunsen burner or alcohol lamp.
☐ A thermometer.

(TO) discover that heat lost equals heat gained for unequal volumes of water. To confirm that kinetic energy is conserved.

WATER MIX (2)   O   Heat (   )

1. Measure out 100 mL of cool water into one of your calibrated styrofoam cups, and record its temperature. Do the same for 50 mL of hot water. Quickly pour the cool into the hot and immediately record the final temperature.

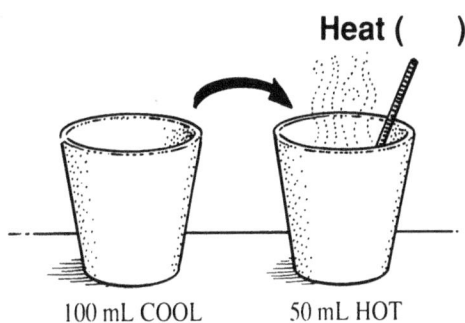

100 mL COOL   50 mL HOT

2. Within the limits of experimental error, does the heat gained by 100 mL of cool water equal the heat lost by 50 mL of hot water? Show all calculations.

3. Repeat your analysis for 50 mL of cool water and 100 mL of hot water.

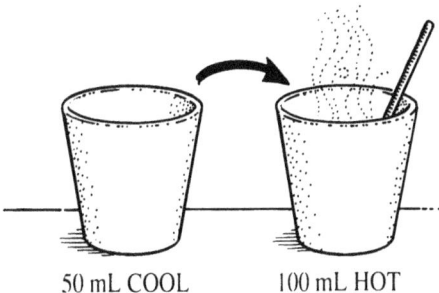

4. This experiment shows that energy is neither created nor destroyed. Explain.

50 mL COOL   100 mL HOT

© 1990 by TOPS Learning Systems   17

Answers / Notes

1-2. Sample data:

1. temperature of cool water = 20.7° C
   temperature of hot water = 43.9° C
   final temperature of mix = 28.0° C

2. Change in cool water:
   28.0° C - 20.7° C = 7.3° C
   Change in hot water:
   43.9° C - 28.0° C = 15.9° C
   Within the limits of experimental error, heat lost = heat gained.

   Heat gained by cool water:
   (100 g) (7.3° C) = 730 calories
   Heat gained by hot water:
   (50 g) (15.9° C) = 795 calories

3. Students should follow a similar procedure as before. Here is one result:

   temperature of cool water = 19.9° C
   temperature of hot water = 46.6° C
   final temperature of mix = 37.4° C

   Change in cool water:
   37.4° C - 19.9° C = 17.5° C
   Change in hot water:
   46.6° C - 37.4° C = 9.2° C
   Again, within the limits of experimental error, heat lost = heat gained.

   Heat gained by cool water:
   (50 g) (17.5° C) = 875 calories
   Heat gained by hot water:
   (100 g) (9.2° C) = 920 calories

4. Heat energy that is lost in one place is always gained some place else. *(Energy can change form, however. For example, a conventional steam engine converts heat energy into mechanical energy. And a nuclear steam reactor converts mass into heat energy.)*

Materials

☐ Graduated styrofoam cups from the previous activity.
☐ Cool and hot tap water, if available. Otherwise, keep heated water in a teapot or ask students to warm their own using a Bunsen burner or alcohol lamp.
☐ A thermometer.

(TO) compare the capacities of water, sand and iron to absorb heat.

HEAT CAPACITY (1)   O   Heat ( )

1. Use a graduated cylinder or a gram balance to fill each styrofoam cup with 75 g of water, sand or iron. All three materials should be at room temperature. Record this temperature.

2. Use your calibrated styrofoam cup to add 75 g of hot water to the first cup. Record temperatures just *before* you pour, and just *after* mixing. Then proceed to the next cup.

75 g HOT

75 g WATER   75 g SAND   75 g IRON

3. Calculate the temperature losses and gains in each cup.

4. Calculate the heat lost by the hot water in each cup. Show your work.

5. Heat lost by the hot water poured into each cup equals heat gained by the water, the sand, and the iron.
   a. Show that this holds true for the cup that contains water.
   b. Which substance has the lowest heat capacity: absorbs the least heat with the greatest rise in temperature? Support your answer with numbers.
   c. Which substances has the highest heat capacity: absorbs the most heat with the smallest rise in temperature? Support your answer with numbers.

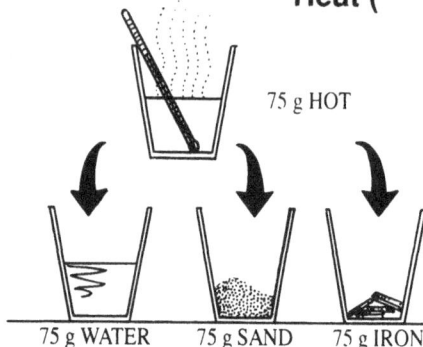

18

Answers / Notes

1-3. Students should add hot water to just one cup at a time, and take all temperature readings before proceeding to the next cup. Here is one result:

	WATER	ΔT	SAND	ΔT	IRON	ΔT
Initial room temperature (°C):	23.0		23.0		23.0	
		31.9		51.0		56.4
Final temperature of mixture (°C):	54.9		74.0		79.4	
		32.6		12.5		6.7
Initial hot-water temperature (°C):	87.5		86.5		86.1	

4.   Heat lost by hot water: (32.6° C)(75 g) =   (12.5° C)(75 g) =   (6.7° C)(75 g) =
                                       2445 calories   938 calories   503 calories

5a. Heat gained by room-temperature water: (31.9° C)(75 g) =
                                            2393 calories

5b. Iron has the lowest heat capacity. It absorbed only 503 calories, yet its temperature rose a full 56.4° C.

5c. Water has the highest heat capacity. It absorbed a full 2,393 calories, yet its temperature rose only 31.9° C.

(This equality cannot be demonstrated for the other 2 cups above, because it has not yet been established by how much the temperature of sand or iron will rise with each additional calorie of absorbed heat.)

Heat lost by each cup of hot water is also gained by the surrounding air (by convection) and the styrofoam cup (by conduction). These gains are negligibly small compared to the heat gained by the contents of each cup.

Materials

☐ Styrofoam cups. One of these should have a 75 ml calibration mark.
☐ Room-temperature water. Fill a suitable container with water and allow it to stand.
☐ Sand.

☐ Iron. Use washers or bolts to make up most of the mass. Add a paper clip or two to fine-tune the mass to 75 grams.
☐ A 100 mL graduated cylinder.
☐ A gram balance and a thermometer.
☐ A calculator (optional).

(TO) compute heat capacities and specific heats for water, sand and iron. To understand how these results influence weather.

HEAT CAPACITY (2)   O   **Heat ( )**

1. The heat capacity of a body is defined as the ratio of calories it absorbs to its change in temperature. Use your results from the previous activity to calculate the heat capacity of water, sand and iron.

$$\text{HEAT CAPACITY} = \frac{\text{calories absorbed}}{\text{change in temperature}} = \text{cal/}^\circ\text{C}$$

2. Specific heat measures the heat capacity of a body divided by its mass.

$$\text{SPECIFIC HEAT} = \frac{\text{heat capacity}}{\text{mass}} = \text{cal/}^\circ\text{C/g}$$

     a. Calculate the specific heat of water, sand and iron.
     b. Does your value for the specific heat of water make sense? Explain.

3. Mix a small pinch of flour in a test tube of water. Angle the tube over a flame and observe the convection currents in the water. Draw what you see.

4. In coastal areas, convection currents cause the wind to blow either onshore or offshore. Keeping specific heats *and* step 3 firmly in mind, which way does the wind blow...

     a. on hot, sunny days? Explain.

     b. on cold nights? Explain.

offshore > < onshore

LAND   SEA

© 1990 by TOPS Learning Systems   19

Answers / Notes

	WATER	SAND	IRON
1. Heat capacity:	$\dfrac{2393 \text{ cal}}{31.9^\circ \text{ C}} = 75.0$ cal/°C	$\dfrac{938 \text{ cal}}{51.0^\circ \text{ C}} = 18.4$ cal/°C	$\dfrac{503 \text{ cal}}{56.4^\circ \text{ C}} = 8.9$ cal/°C
2a. Specific heat:	$\dfrac{75.0 \text{ cal/}^\circ \text{ C}}{75 \text{ g}} = 1.00$ cal/°C/g	$\dfrac{18.4 \text{ cal/}^\circ \text{ C}}{75 \text{ g}} = 0.25$ cal/°C/g	$\dfrac{8.9 \text{ cal/}^\circ \text{ C}}{75 \text{ g}} = 0.12$ cal/°C/g

2b. Yes. The specific heat of water is 1.00 cal/°C/g. This is exactly how one calorie is defined: the amount of heat required to raise 1 gram of water 1 °C.

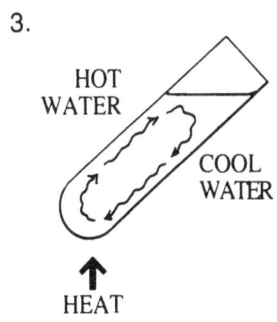

3.

HOT WATER

COOL WATER

↑ HEAT

4a. On hot, sunny days the wind tends to blow onshore. Even though the sun's rays fall equally on land and sea, the land (equivalent to sand) heats to a higher temperature than the sea (equivalent to water) because it has a lower specific heat. As a consequence, the air over the hotter land expands and rises. Denser air over the cooler sea flows in to take its place, setting up on-shore convection currents similar to the currents modeled in the test tube.

4b. On cold nights the wind tends to blow offshore. The land, with its lower heat capacity, cools more rapidly. The sea, with its higher heat capacity, retains more heat longer, eventually warming its air more than the land. And so the sea air begins to rise while cooler land air flows in to take its place, setting up off-shore convection currents.

Materials

☐ A calculator.
☐ A pinch of flour.
☐ A test tube.
☐ A candle, Bunsen burner, or alcohol lamp, plus matches.

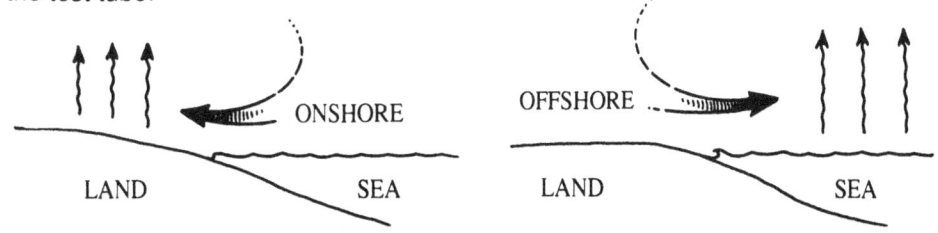

ONSHORE

LAND   SEA

OFFSHORE

LAND   SEA

(TO) experimentally determine the heat content of a peanut.

PEANUT POWER   ◯   Heat (   )

1. Find the mass of a half peanut on a gram balance.

2. Push a straight pin through masking tape and press it to a tin can lid so the pin points upright. Gently stick your half peanut firmly on this point.

3. Invert a pie tin over 3 small jars. Fill another pie tin with 500 mL of water and rest it on top. Record the temperature of the water to .1° C.

4. Light the half peanut with a match just under the edge of the pan. As soon as it is vigorously burning, slide it to the center.

500 mL WATER

PEANUT HALF

LID

  a. Record your observations. Did the water absorb *all* the heat?

  b. Record the highest temperature reached by the water to .1° C.

  c. Find the mass of the charred remains of your peanut. Calculate the calories of heat per gram that were generated by the part that burned.

  d. Calculate the calories generated per gram of peanuts from information listed on the package. Remember that 1 Calorie = 1000 calories.

  e. How efficiently did your experimental apparatus trap heat?

© 1990 by TOPS Learning Systems   20

Introduction

If you're on a diet, you'll want to stay away from potato chips. Consider how many Calories (kilocalories) you consume by eating just one. And who can stop with just one! (Stick a potato chip to a pin as in step 2. It won't break if you gently drill it in. Now light it with a match. Be prepared to ventilate your room after this miniconflagration.)

This fire doesn't extract energy from the chip in exactly the same manner as your body. But the amount of energy is the same in both cases. How would you measure this energy? (Use it to heat water. Measure the mass of the water, its temperature increase, then multiply to find the number of calories absorbed.)

Can you think of a way to trap 100% of the heat energy given off by the potato chip? (In actual practice, it is fully dehydrated and reduced to powder, placed in a hollow steel cylinder, sealed with a screw cap, and totally submerged in water. The contents are then pressurized with oxygen through appropriate fittings and ignited by an electrical spark to produce a fully contained explosion. *All* heat energy from this explosion is absorbed by the surrounding water.)

Answers / Notes

1-4. *Here is a model energy analysis for the half peanut:* The half peanut burns vigorously, after a slow start, to a black husk. Some heat escapes the water during this process: smoky, warm air fills the cavity under the pie plate, seeps out from the perimeter, and convects upward. Heat is also radiated directly into space. *(Some heat is also used to heat up the pie plates. Because of the low specific heat of the aluminum, this is small compared to the total mass of the water.)*

		calories absorbed by water:	$(1.6° C)(500 g)$ = 800 calories
mass of unburned peanut:	.49 g		
mass of burned peanut:	.06 g	experimental heat content:	$\dfrac{800 \text{ calories}}{.43 \text{ g}}$ = 1,860 cal./g
mass of fuel consumed:	.43 g		
final temperature of water:	23.1° C	calorie information on package:	$\dfrac{170,000 \text{ calories}}{28.4 \text{ g}}$ = 6,000 cal./g
initial temperature of water:	21.5° C		
change in temperature:	1.6° C	efficiency:	$\dfrac{1,860 \text{ cal./g}}{6,000 \text{ cal./g}}$ x 100 % = 31 %

Materials

☐ A bag or can of roasted peanuts. (Read **basic safety rule: ALLERGIES**, page C.) Calories per serving should be clearly labeled for student use: one ounce (28.4 g) contain 170 Calories, or 170,000 calories.

☐ A gram balance, thermometer, and hand calculator.

☐ A straight pin, masking tape, tin can lid and matches.

☐ Three baby food jars or small beakers, and two disposable aluminum pie tins.

☐ A 100 mL graduated cylinder. Use the 500 mL calibrated jars from activity 15, if available.

REPRODUCIBLE
STUDENT
TASK CARDS

Task Cards Options

Here are 3 management options to consider before you photocopy:

1. Consumable Worksheets: Copy 1 complete set of task card pages. Cut out each card and fix it to a separate sheet of boldly lined paper. Duplicate a class set of each worksheet master you have made, 1 per student. Direct students to follow the task card instructions at the top of each page, then respond to questions in the lined space underneath.

2. Nonconsumable Reference Booklets: Copy and collate the 2-up task card pages in sequence. Make perhaps half as many sets as the students who will use them. Staple each set in the upper left corner, both front and back to prevent the outside pages from working loose. Tell students that these task card booklets are for reference only. They should use them as they would any textbook, responding to questions on their own papers, returning them unmarked and in good shape at the end of the module.

3. Nonconsumable Task Cards: Copy several sets of task card pages. Laminate them, if you wish, for extra durability, then cut out each card to display in your room. You might pin cards to bulletin boards; or punch out the holes and hang them from wall hooks (you can fashion hooks from paper clips and tape these to the wall); or fix cards to cereal boxes with paper fasteners, 4 to a box; or keep cards on designated reference tables. The important thing is to provide enough task card reference points about your classroom to avoid a jam of too many students at any one location. Two or 3 task card sets should accommodate everyone, since different students will use different cards at different times.

HOT WIRE ◯ Heat (   )

1. Cut a piece of iron wire about 10 cm long (as tall as this task card). 2. *Partly* melt a small lump of wax on a microscope slide. Roll the wire in this wax, just as it begins to resolidify, to thoroughly coat it with wax.

3. Hold the end of the wire just above the flame so it tilts slightly down. Describe how heat conducts (travels) from the flame through the wire.

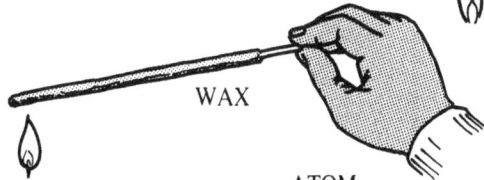

WAX

4. Atoms in the wire vibrate more vigorously as they absorb heat energy. Propose a theory to explains how agitated atoms near the flame pass this energy along to atoms in the middle of the wire.

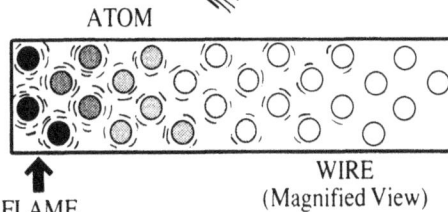

ATOM

FLAME

WIRE
(Magnified View)

5. Gases in a candle flame ignite at about 600° C. Wax liquefies around 60° C. Temperatures above 40° C feel uncomfortably hot. Describe the temperatures present in your hot wire.

1

HEAT RACE ◯ Heat (   )

1. Coat 3 pieces of wire with wax as before. Each should be about 10 cm long, the height of this card.

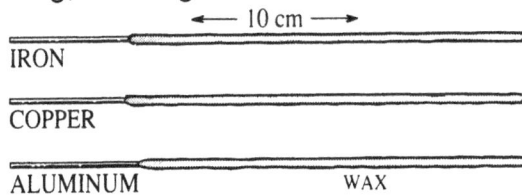

←— 10 cm —→

IRON

COPPER

ALUMINUM   WAX

2. Hold heat races: Place 2 wires side by side, tipped slightly down, over a candle flame. Summarize your results in a table.

Race	Result
iron vs. copper	
iron vs. aluminum	
copper vs. aluminum	

3. Heat is conducted through metals because atoms vibrate against each other. Even more importantly, their free outer electrons also collide with each other, and other atoms.

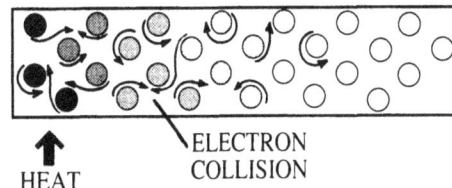

HEAT

ELECTRON
COLLISION

  a. Which metal likely holds its outer electrons most tightly? What makes you think so?

  b. Would you expect good conductors of heat to be good conductors of electricity (electrons) as well? Explain.

2

CONDUCTORS / INSULATORS ○   Heat ( )

1. Cut a strip of aluminum from the bottom of a pie plate, equal in size to a microscope slide.

2. Coat 1 surface of each with melted wax, and let them cool.

3. Hold another heat race: place a corner of both "slides" side by side over a candle flame. Write your observations in words and pictures.

4. Substances that slow the passage of heat are called insulators. Which is the insulator and which is the conductor?

5. Wax that gets very hot vaporizes to a gas. Did any parts of either slide get this hot? Why?

6. Metals have free outer electrons. Do you think this is true for nonmetals like glass?

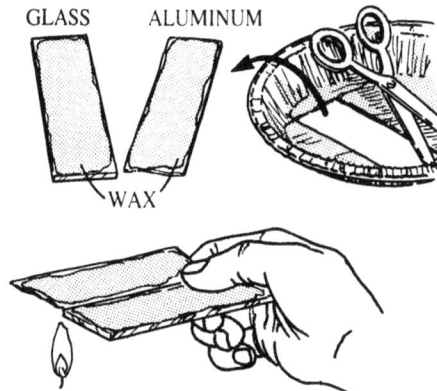

3

COLD FINGER   ○   Heat ( )

1. Set an ice cube in a pie tin and touch it with your finger.
   a. Explain why your finger begins to feel cold. (Hint: Heat travels, not cold. Cold is merely the absence of heat energy.)
   b. How long could heat continue to flow in this experiment? Think carefully!

2. Pinch the ice cube between a glass microscope slide and the aluminum slide you made previously. Does one side feel colder than the other? Explain.

3. Puff up a plastic sandwich bag somewhat, then seal it air-tight with a strip of masking tape.

4. Wrap your ice cube in the baggy, cushioned by air.
   a. Evaluate air as an insulating material.
   b. Compare the insulating properties of air with plastic.

5. Do you think a vacuum (nothing at all) would be a good insulator? Why do you think so?

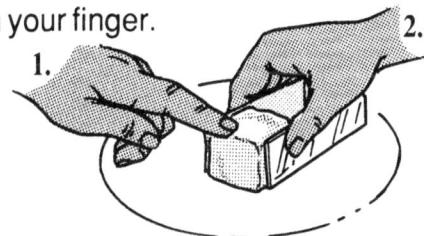

4

RISE AND FALL   O   Heat ( )

1. Mix 1 drop of blue food coloring in a jar of cold water. Fill a smaller jar brim full, using a pie plate to catch any overflow.

COOL BLUE WATER

2. Fill a second jar with hot water, brim full like the first. Keep both in the same overflow container.

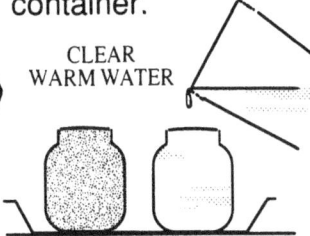

CLEAR WARM WATER

3. Cover the cool blue water with an index card, and carefully invert it over the warm clear water. Carefully pull out the card and write your observations.

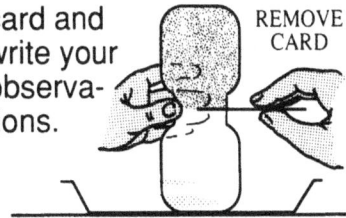

REMOVE CARD

4. Repeat the experiment. This time invert the hot clear water over the cool blue water. Write your observations.

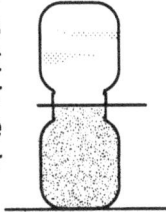

5. Warm water rises as cool water sinks. Heat movement of this kind is called *convection*.

    a. Did you observe convection in steps 3 and 4? Explain.

    b. Do atoms *conduct* heat through a wire in the same manner as molecules *convect* heat through a fluid? Explain.

© 1990 by TOPS Learning Systems   5

CONVECTION MACHINE   O   Heat ( )

1. Cut out a coil from an index card.

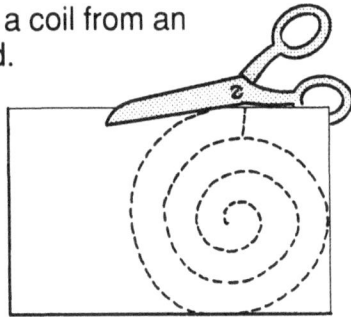

2. Tape a thread to the center of your coil, then suspend it at least a hand span above a burning candle. Be careful it doesn't catch fire.

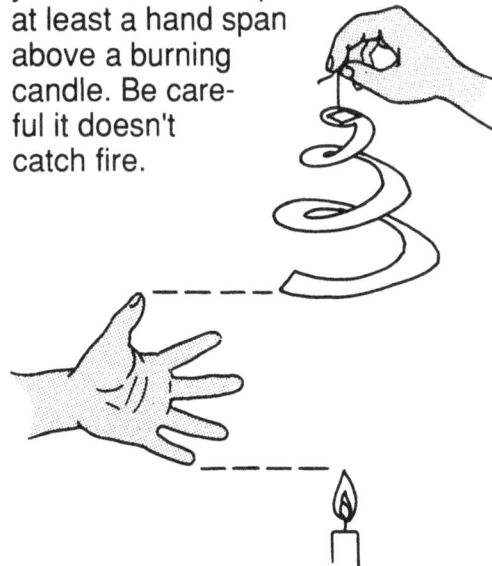

3. Tell how your convection machine works. Illustrate your answer with a diagram.

4. What makes heated air rise?

© 1990 by TOPS Learning Systems   6

TOO HOT TO HANDLE?   O   Heat (   )

1. Fill a test tube 1/5 full of crushed ice. Push a small plug of steel wool over the ice with your pencil, then fill the tube with water.

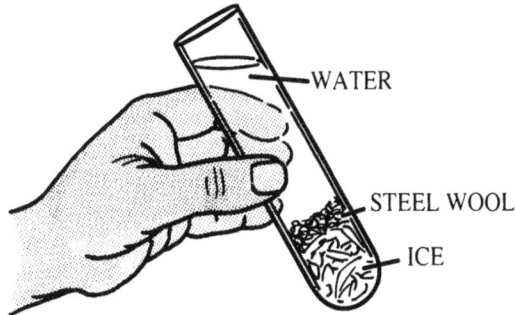

2. Try to boil water at the top of the tube without melting the ice at the bottom. Is water a conductor of heat, or an insulator? Explain.

WATER

STEEL WOOL

ICE

3. Would this experiment turn out the same with ice at the top of the tube and the flame at the bottom? Why?

7

PAPER COOKING POT   O   Heat (   )

1. Press a scrap of notebook paper into the mouth of a test tube with your thumb.

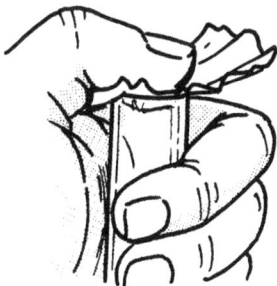

2. Cut it into a spoon shape. Slip a paper clip over the handle so its end supports the bowl.

3. Trim off any excess paper, then attach a clothespin. Fill with water and heat it over a candle flame. Write your observations.

SUPPORT THE BOWL

4. Water boils at 100° C (at sea level). It will not get hotter than this until all the liquid is vaporized to steam.
   a. Is the combustion temperature of paper higher than 100° C? Explain.
   b. Will the spoon eventually burn if you leave it over the flame? Explain.

5. Explain the role of conduction and convection in keeping the paper cool.

8

RADIATION   O   Heat (   )

1. Probe the edge of a candle flame with wax shavings stuck to the end of a toothpick. Sketch the *isotherm* (an equal-temperature line) surrounding the flame where this wax just starts to melt (about 60° C).

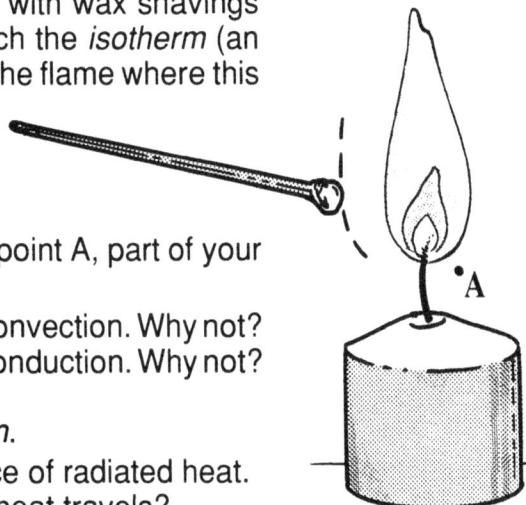

2. Suppose candle wax just melts at point A, part of your 60° C isotherm.
   a. Heat cannot travel to point A by convection. Why not?
   b. Heat cannot travel to point A by conduction. Why not?

3. Heat travels to point A by *radiation*.
   a. Name our most important source of radiated heat.
   b. How fast do you think radiated heat travels?

4. Thoroughly heat an iron wire in the hot flame of an alcohol lamp or Bunsen burner. What evidence can you see that radiated heat energy is similar to light?

9

REFLECTION/ABSORPTION   O   Heat (   )

1. Clamp clothespins onto 2 tin can lids. Cover just 1 side of 1 lid with candle soot; leave the other lid shiny on both sides.

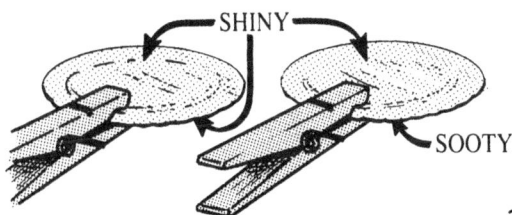

2. Drip melted candle wax in the middle of each lid (*not* on the blackened side). Quickly stick a penny to each lid before the wax dries.

SHINY

PENNY

SOOTY

3. Hold each lid so the pennies are directly opposite the candle flame, behind each lid. Keep them a small but equal distance away.
   a. Which penny drops first?
   b. Explain your observations in terms of absorption and reflection.

SHINY   SOOTY

4. Light is absorbed and reflected like radiated heat. Explain why this makes sense.

10

BEST EMITTER?　　　O　　　Heat (　)

1. Tightly wrap a clean, dry test tube in each material. Use just enough to fully cover the glass. Hold the wrapper with clear tape.

BLACK PAPER
WHITE PAPER
ALUMINUM FOIL

POUR SPOUT

2. Stand the tubes upright in a rack (a jar divided by rubber bands will serve). Pinch a pour spout into a styrofoam cup. Use it to fill each tube with boiling water, while keeping the outside of each tube *perfectly dry*.

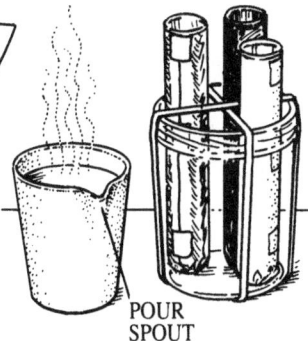

3. Wait 10 minutes or longer for the test tubes to cool. While you are waiting...
   a. *Briefly* touch the side of each tube. Which feels hottest? Why?
   b. Predict which tube will be the best heat emitter (will cool the fastest).

4. After 10 minutes or so, use a thermometer to quickly measure the temperature of the water in each tube. Do this in the same order you filled them.

5. Was your prediction correct? Propose a theory to explain your observations. (Seal your filled test tubes with corks or clay. Save them for the next activity.)

11

BEST ABSORBER?　　　O　　　Heat (　)

1. Get your stoppered test tubes full of water from the previous activity. They should all be at room temperature.

2. Expose these test tubes to about 10 minutes of direct sunlight outdoors, or use a heat lamp or high wattage light bulb indoors. Slant the tubes to receive perpendicular rays.

3. While you are waiting, predict which tube should best absorb heat (warm the fastest). Recall results from previous activities to support your answer.

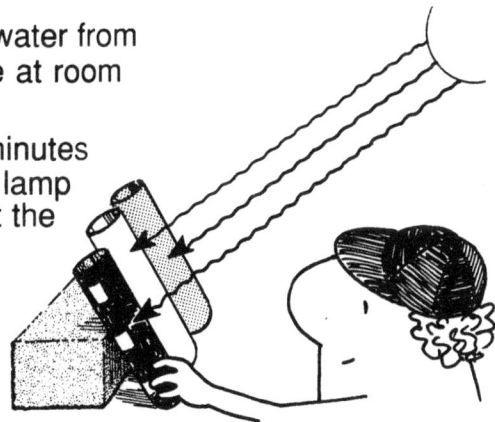

4. After about 10 minutes, quickly measure the final temperature of the water in each tube, while shielding all tubes from additional radiation.
   a. Evaluate your prediction. Was the best absorber also the best emitter?
   b. What would happen if a good absorber were a poor emitter?

5. When does a test tube emit heat radiation, and when does it absorb it?

6. What color should you paint your house to conserve energy? Explain.

12

THE GREENHOUSE EFFECT   O   Heat ( )

Fact A: The sun (any hot body) radiates higher-energy short waves. The earth (any cool body) radiates lower-energy long waves.

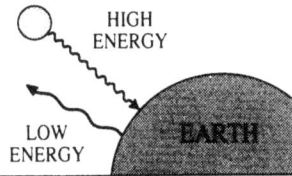

HIGH ENERGY

LOW ENERGY   EARTH

Fact B: Glass transmits shorter light waves, but not longer infra-red waves.

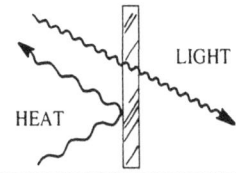

LIGHT

HEAT

1. Fill 2 small jars about 1/6 full of sand and just cover with water. Cap one with a third jar and seal with masking tape. Leave the other uncovered.

    a. Will the wet sand in each jar warm equally if you set both jars in the sun? Apply facts **A** and **B** as you answer.

    b. Test your prediction for 10 minutes (or longer) and report your results.

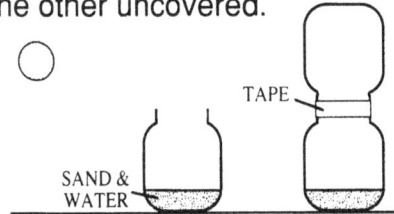

TAPE

SAND & WATER

2. Carbon dioxide in Earth's atmosphere is transparent to shorter wave lengths and absorbs longer wavelengths, in a manner similar to glass.

    a. Explain why it is important to reduce our use of gasoline, coal and oil.

    b. What can *you* do to make a difference?

EARTH

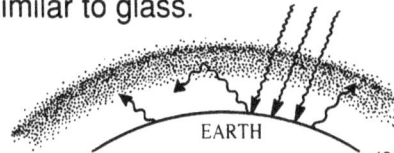

© 1990 by TOPS Learning Systems   13

HOLD THAT HEAT   O   Heat ( )

1. Mark where 100 mL of water reaches inside a styrofoam cup.

2. Fill it to this mark with near-boiling water, then pour it into a metal can. Record the highest temperature reached while you note the position of a second hand on your watch or a wall clock.

3. Mark the temperature of the water after each new minute for the next 10 minutes. Stir the water once or twice before reading each temperature.

4. Repeat this experiment for a glass jar. Present your time and temperature data in a table, and graph your results.

5. Interpret your data.

6. Which forms of heat loss made the most difference in this experiment? Explain why you think so.

time (min)	temperature (°C)	
	metal can	glass jar
0		
1		
2		
3		
⋮		
10		

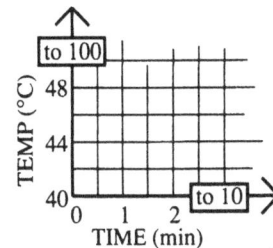

to 100

48

44

40

TEMP (°C)

0   1   2   to 10

TIME (min)

© 1990 by TOPS Learning Systems   14

COLD HANDS   O   Heat (   )

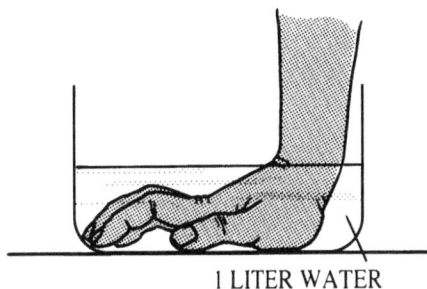

1 LITER WATER

1. Add 1000 g of room-temperature water to a large container. Accurately record its temperature to the nearest .1° C.

2. Completely submerge your *right* hand in this water for exactly 5 minutes. (Count each minute so you don't lose track.) Accurately record the final temperature and calculate the increase.

   a. One *calorie* (a unit of energy) raises 1 gram of water 1° C. How many calories of heat are required to raise the water in your container 1° C?
   b. How many calories did your hand transfer to the water?

3. Assume your *left* hand gives off the same amount of heat as your right.
   a. Predict how much your left hand will raise the temperature of 500 g of water in 5 minutes. Show your calculations.
   b. Test your prediction. (Keep your *whole* hand under water as before.)

4. Does a thermometer measure temperature or heat? Refer to your results in steps 2 and 3 as you answer.

15

WATER MIX (1)   O   Heat (   )

1. Use a graduated cylinder to accurately calibrate the inside of 2 styrofoam cups. Mark the 50 mL, 75 mL and 100 mL levels.

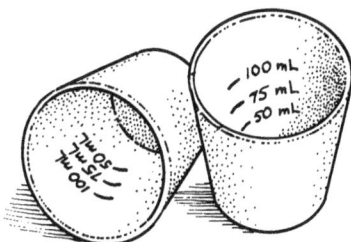

2. Measure 75 mL of cool water into one cup and record its temperature. Do the same for 75 mL of hot water. Quickly pour the cool into the hot and immediately record the final temperature.

75 mL COOL

75 mL HOT

3. Calculate the change in temperature for each water sample.

4. Compute the number of calories gained by the cool water and lost by the hot water.

5. Within the limits of experimental error, does the heat gained by the cool water equal the heat lost by the hot water? What are some sources of error?

(Save your graduated cups.)

16

WATER MIX (2)　　　　O　　　　Heat (　)

1. Measure out 100 mL of cool water into one of your calibrated styrofoam cups, and record its temperature. Do the same for 50 mL of hot water. Quickly pour the cool into the hot and immediately record the final temperature.

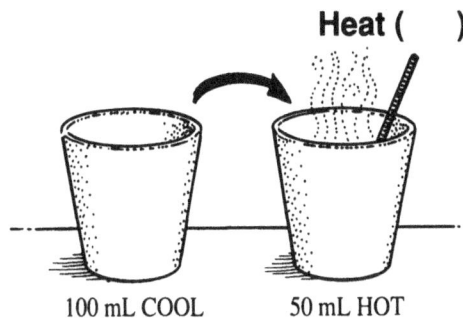

100 mL COOL　　　50 mL HOT

2. Within the limits of experimental error, does the heat gained by 100 mL of cool water equal the heat lost by 50 mL of hot water? Show all calculations.

3. Repeat your analysis for 50 mL of cool water and 100 mL of hot water.

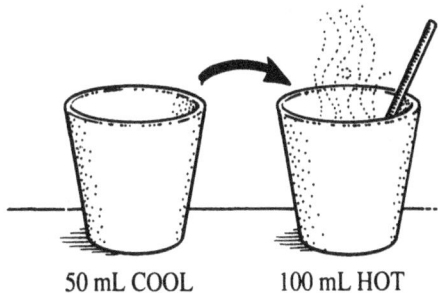

4. This experiment shows that energy is neither created nor destroyed. Explain.

50 mL COOL　　　100 mL HOT

17

HEAT CAPACITY (1)　　　　O　　　　Heat (　)

75 g HOT

1. Use a graduated cylinder or a gram balance to fill each styrofoam cup with 75 g of water, sand or iron. All three materials should be at room temperature. Record this temperature.

2. Use your calibrated styrofoam cup to add 75 g of hot water to the first cup. Record temperatures just *before* you pour, and just *after* mixing. Then proceed to the next cup.

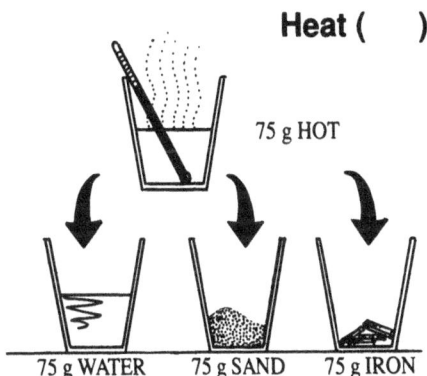

75 g WATER　　75 g SAND　　75 g IRON

3. Calculate the temperature losses and gains in each cup.

4. Calculate the heat lost by the hot water in each cup. Show your work.

5. Heat lost by the hot water poured into each cup equals heat gained by the water, the sand, and the iron.
    a. Show that this holds true for the cup that contains water.
    b. Which substance has the lowest heat capacity: absorbs the least heat with the greatest rise in temperature? Support your answer with numbers.
    c. Which substances has the highest heat capacity: absorbs the most heat with the smallest rise in temperature? Support your answer with numbers.

18

HEAT CAPACITY (2)   O   Heat ( )

1. The heat capacity of a body is defined as the ratio of calories it absorbs to its change in temperature. Use your results from the previous activity to calculate the heat capacity of water, sand and iron.

$$\textbf{HEAT CAPACITY} = \frac{\textbf{calories absorbed}}{\textbf{change in temperature}} = \textbf{cal/}°\textbf{C}$$

2. Specific heat measures the heat capacity of a body divided by its mass.

$$\textbf{SPECIFIC HEAT} = \frac{\textbf{heat capacity}}{\textbf{mass}} = \textbf{cal/}°\textbf{C/g}$$

   a. Calculate the specific heat of water, sand and iron.
   b. Does your value for the specific heat of water make sense? Explain.

3. Mix a small pinch of flour in a test tube of water. Angle the tube over a flame and observe the convection currents in the water. Draw what you see.

4. In coastal areas, convection currents cause the wind to blow either onshore or offshore. Keeping specific heats *and* step 3 firmly in mind, which way does the wind blow...

   a. on hot, sunny days? Explain.

   b. on cold nights? Explain.

offshore ⟩ ⟨ onshore

LAND   SEA

19

PEANUT POWER   O   Heat ( )

1. Find the mass of a half peanut on a gram balance.

500 mL WATER

2. Push a straight pin through masking tape and press it to a tin can lid so the pin points upright. Gently stick your half peanut firmly on this point.

3. Invert a pie tin over 3 small jars. Fill another pie tin with 500 mL of water and rest it on top. Record the temperature of the water to .1° C.

4. Light the half peanut with a match just under the edge of the pan. As soon as it is vigorously burning, slide it to the center.

PEANUT HALF   LID

   a. Record your observations. Did the water absorb *all* the heat?

   b. Record the highest temperature reached by the water to .1° C.

   c. Find the mass of the charred remains of your peanut. Calculate the calories of heat per gram that were generated by the part that burned.

   d. Calculate the calories generated per gram of peanuts from information listed on the package. Remember that 1 Calorie = 1000 calories.

   e. How efficiently did your experimental apparatus trap heat?

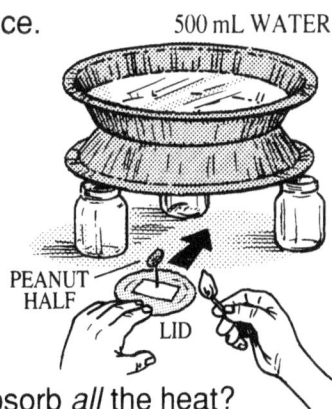

20

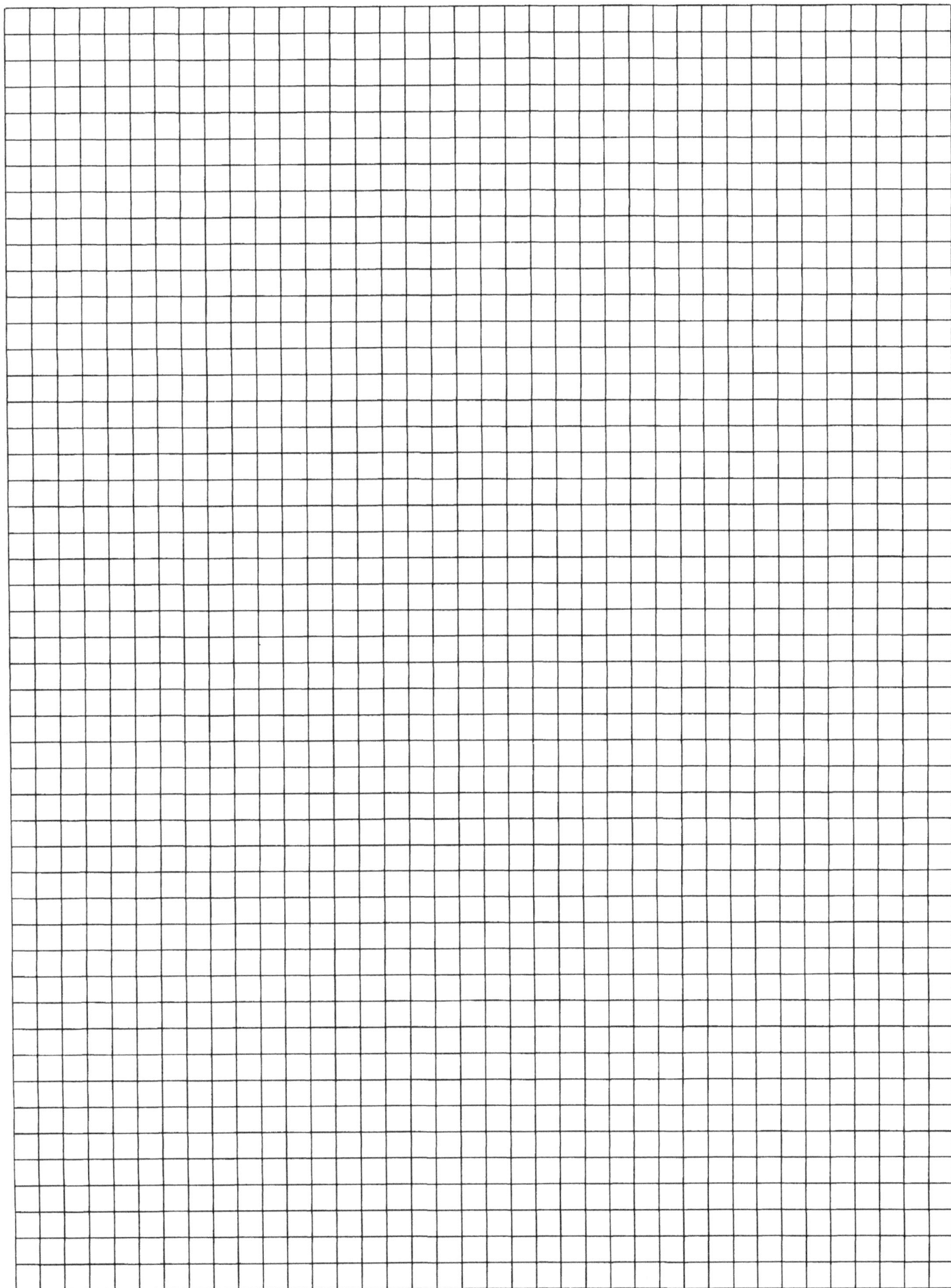

Feedback

If you enjoyed teaching TOPS please tell us so. Your praise motivates us to work hard. If you found an error or can suggest ways to improve this module, we need to hear about that too. Your criticism will help us improve our next new edition. Would you like information about our other publications? Ask us to send you our latest catalog free of charge.

For whatever reason, we'd love to hear from you. We include this self-mailer for your convenience.

Sincerely,

Ron and Peg Marson
author and illustrator

Your Message Here:

Module Title _____ Date _____

Name _____ School _____

Address _____

City _____ State _____ Zip _____

——————————————— FIRST FOLD ———————————————

——————————————— SECOND FOLD ———————————————

RETURN ADDRESS

_____

_____

_____

PLACE
STAMP
HERE

TOPS Learning Systems
342 S Plumas St
Willows, CA 95988

TAPE HERE